POWER GRAB:

The Conserver Cult and the Coming Energy Catastrophe

POWER GRAB:

The Conserver Cult and the Coming Energy Catastrophe

James A. Weber

ARLINGTON HOUSE·PUBLISHERS
165 HUGUENOT STREET · NEW ROCHELLE, NEW YORK 10801

Manufactured in the United States of America
P 10 9 8 7 6 5 4 3 2 1

Library of Congress Cataloging in Publication Data

Weber, James A
    Power grab.

    Includes index.
    1. Power resources. 2. Environmental protection.
I. Title.
TJ163.2.W37   333.7   79-21583
ISBN 0-87000-453-0

CONTENTS

To my wife, Shirley

PREFACE

How often has public calamity been arrested on the very brink of ruin, by the seasonable energy of a single man?

EDMUND BURKE

While writing this book, I happened to live through a big-city blizzard. Chicago's metropolitan area was buried in several feet of snow and frozen by more than a week of zero and below temperatures. Everything was immobilized. Car traffic came to a halt. Buses and trains didn't run. Because of snowpacked streets, ambulances experienced great difficulties getting to people requiring medical help. Firemen were reduced to using toboggans borrowed from local parks to transport hoses through the streets because fire engines could not get through the snow. Several weeks of garbage backed up in alleys and on curbs. People fell off their roofs while shoveling snow, breaking arms and legs. Heart attacks and other medical problems aggravated by the adverse weather conditions

7

took their toll. It was reported that businesses lost $1 billion in sales and employees $200 million in wages. Aside from the continuing aggravation that it brought in its wake, the blizzard made me think of two things. One was that it was a taste of what a full-scale energy crisis would be like—with one big difference. During the blizzard and its aftermath, we still had heat and light in our homes and communications in the form of radio, television, and telephone. We also had energy to combat, remove, and eventually overcome the snow. By contrast, in an energy crisis, with shortages of oil and gas as well as an electricity blackout, the lights would go out, there would be no heat, and communications would be shut down. The problems of the blizzard would be small by comparison.

My other thought was that the blizzard was brought to us by none other than the ever-popular "environment." Yes, the environment. In recent years, we have all heard—over and over and over again—of the many apparently horrible things that we have been doing to our environment. We have heard very little of what the environment can and does do to us. But the fact that nearly 100 people were killed by the blizzard mentioned above is a reminder that the environment is not completely sustaining, friendly, or helpless. It can, in fact, be a most powerful, implacable and destructive force.

We are sometimes told that because most Americans live in cities, we have been cut off from the environment and therefore are not aware of the terrible things we do to it. But it works the other way too. Because most of us live in urban areas with all of the protections and conveniences provided by these areas in terms of housing, heating, lighting, and other services, we tend to forget what the environment can do *to us*. The blizzard helped jog my memory.

We, of course, live in the environment. But it also should be obvious that we must protect ourselves from

8

and in effect control the environment if we are not only to survive but prosper. Human advancement has in fact occurred because of our increasing ability to control our environment rather than be controlled by it. The primary, the essential tool we have used to put the environment to work for us is energy: human energy, our arms and legs and backs; animal energy, provided by oxen, mules and horses; kinetic energy, produced by the wind blowing and water flowing; chemical energy, released as fire by burning wood, coal, oil, gas; and nuclear energy in the form of fission. In the future there will be geothermal, fusion, solar, hydrogen, and who knows what other energy forms now conceived of only dimly if at all.

Energy is to human society as blood is to the human body. However, there is today a movement in America and abroad which believes that growing energy production and use is deleterious because of its adverse effects, alleged and real, on the environment and mankind. This movement proposes to practice "energyletting" to "cure" us of our environmental and social ills, much like physicians of old practiced bloodletting to eliminate diseases of the human body. But more patients died from bloodletting than from the diseases it was supposed to cure. The same will be true for human society as a result of energyletting.

Are we to control the environment and maintain and expand human society through the growing use of energy, or is the environment to be used as a pretext for controlling society and eventually decimating humanity for lack of energy? The answer to this question is the subject of this book. It is an answer that, paradoxically, was made plainer in the midst of a blizzard.

My deepest thanks go to a most understanding editor, Karl Pflock, and an equally understanding typist, Carrie Steif. Their complete professionalism made this book immeasurably easier to write.

1

WHAT HAPPENED IN 1970?

Industry is where it strikes first. People in manufacturing jobs are laid off in thousands, then hundreds of thousands, finally millions. Next to go on the unemployment lists are double to triple the number of people in service jobs. The heaviest job losses are in urbanized areas where the bulk of employment is concentrated.

Now, transportation systems begin to falter and fail. Railroads and trucks run less frequently or not at all, resulting in shortages in food and other commodities. Subways and buses shut down, eliminating mass transportation. Cars are used only for short trips or sit idle. People are left stranded with no means of transportation and dwindling supplies of goods.

Homes go without heat or air conditioning, first on a spot basis, then a growing national basis. Appliances no longer operate, resulting in food spoilage due to refrigerator failures and lack of cooking capacity. Living stresses increase and there are growing medical problems, particularly among old people with physical disabilities and children with low resistance to disease.

Communications become increasingly inoperative. Television and radio stations stop broadcasting, telephones don't work, newspapers are shut down. Offices and stores close. Street and home lights go out.

Thrown out of jobs, cut off from communications, plunged into darkness, faced with growing shortages, people panic. Mobs gather in the streets, rioting and looting erupt, violence against people and property become the order of the day.

It is not a pretty picture. Indeed, it is a gruesome picture. Yet it is a real picture of the energy catastrophe towards which this country is heading unless positive and rapid steps are taken to increase domestic energy production.

Listen to Dr. John J. McKetta, director of the National Council for Environmental Balance and professor of chemical engineering at the University of Texas. Formerly chairman of the Advisory Committee on Energy to the Secretary of the Interior and chairman of the Committee on National Air Quality Management for the National Academy of Science, McKetta warns that "the United States is now facing disastrous consequences from an inability to meet energy needs. Since the gravity of our problem was spotlighted by the Arab oil embargo in 1973, almost everything we have done has tended to worsen the situation. Time is running out. Even if we could move tomorrow to develop a workable energy policy and were able to implement it at once, we would still experience at least a deterioration in our standard of living in this country by 1985. If we do not move quickly, the situation will become direct. I mean that we will experience deep and painful disruptions in our whole economy. We will have a severe recession by 1985 brought about by shortages of domestic energy. In fact, unless we move immediately, there will be an energy shortage in the United States that we simply cannot imagine at this time."[1]

Electricity is a case in point. For the past several years, electric utility spokesmen have been routinely warning of impending shortages. In 1976, for example, the National Electric Reliability Council (NERC) stated that the United States had moved closer to a "severe electric energy crisis" during the previous year and could face blackouts or power reductions in the eighties.[2]

In 1977, NERC stated that power blackouts and government-imposed restrictions on the use of electricity were likely to be common in some parts of the country by 1979 and in most areas by 1986. "We've been asked what could be done to minimize the problem of these shortages occurring as early as we say they might," said NERC chairman C. E. Winn. "The fact is that there's very damn little that could be done in the short term." Norton Savage, chief of power supply for the Federal Power Commission, said the FPC agreed with the industry report. "The only real difference between us and them is that they see problems beginning in 1979," Savage said. "We don't see any problems until 1981."[3]

In 1978, NERC reiterated that the United States faced the "grim prospect" of power shortages in the early 1980s. According to NERC, the long-term outlook for adequacy and reliability of the U.S. power supply became "materially worse" during the year, even though utility company forecasts of future electricity load growth were lower than those previously projected. Shortages were expected to develop locally, then regionally, and, finally, to spread throughout the United States.[4]

Coal, the primary fuel for electricity production, has been touted as the energy superman that will save the day. In the past, optimistic forecasts have been made that coal production will double by 1985. But there is no way that this can happen under present production conditions. To double by 1985, coal production would have to increase by an average of more than 10 percent a year. But, in 1977 when electricity usage increased by 8 percent, coal production increased by only 1.5 percent.[5]

Nuclear power today is an increasingly more important source of electricity. But, rather than gearing up to meet electrical demand, the nuclear industry is being wound down as shown by letters of intent. In 1977, the nuclear industry filed only four letters of intent to build new nuclear plants. In 1978, the industry filed only two, and these barely slipped in eleven days before the end of the year.[6]

Oil and gas are primarily used for transportation, heating, and industrial processing. The domestic oil supply has flattened out in recent years despite the obvious need to increase production in order to reduce imported oil to a minimum. It was back in 1973 that OPEC (the Organization of Petroleum Exporting Countries) embargoed oil and began increasing prices. Yet, since then, oil imports have grown from 34 to close to 50 percent of consumption while domestic oil production has stagnated at 8 to 9 million barrels a day. At the same time, production of natural gas has leveled out at about 20 trillion cubic feet a year. Although bright forecasts are made for both of these fuels, neither at this time shows signs of yielding substantial production increases in the future.

These four energy sources—coal, nuclear, oil and gas—currently provide about 96 percent of our domestic energy supply, with hydroelectric accounting for the balance. However, the picture is not one of energy sources dynamically growing to meet increasing demand but of an energy supply that is static and faltering, resulting in growing energy shortages. These shortages can only be met by importing more costly energy in the form of foreign oil and liquefied natural gas. But this increases costs while exposing Americans to the possibility of catastrophic shortfalls brought on by unavailability of energy supplies or steeply rising energy costs or both. At the same time, it creates the possibility of an over-whelmingly dangerous international confrontation at a time when the Soviet Union is rapidly extending its influence and power in the Persian Gulf, our major source of foreign oil.

There is no way that Americans desire these results. Huge majorities of people, in fact, recognize the vast importance of energy and strongly favor rapid development of all available energy resources in order to prevent shortages while minimizing dependence on imported supplies. According to Louis Harris, who polled the public in mid-1978, a majority of 94 to 2 percent want "work on solar energy expanded." By 89 to 7 percent, people "want more drilling for oil and natural gas." An 87 to 7 percent majority favors "building more plants that can convert coal to natural gas and oil." By 83 to 10 percent, people support "expanding underground mining for coal." By 78 to 15 percent, a majority would like to see a "speed up in the drilling for oil and natural gas off the Atlantic coast," and a 76 to 16 percent majority supports "allowing more drilling off the Pacific and Gulf coasts." By 76 to 12 percent, people want "the time required to get government approval to construct major energy facilities reduced." By 73 to 8 percent, they favor "converting oil shale to synthetic crude oil." By 73 to 15 percent, people would like to have "more coal-fired electric-generating plants built." By 69 to 22 percent, a majority favors "building more nuclear power plants to generate electricity." (The pronuclear percentage dropped but still remained a majority after the Three Mile Island incident.) By 68 to 21 percent, most stand in favor of "providing economic incentives for oil and gas companies to explore and develop natural resources." By 61 to 26 percent, a majority favors "expanding surface mining for coal."[7] In addition, by 65 to 23 percent, a majority of Americans favors "deregulation of the price of all oil produced in the United States, if this would encourage development of more oil production here at home."[8] At the same time, by 67 to 24 percent, people oppose "importing more oil and liquefied gas from overseas."[9]

The 1973 oil embargo clearly alerted the country to the need for action to avert future energy shortages. There are basically two things that can be done. One is to produce more energy at home. The other is to conserve energy either by using it more efficiently or by simply using less of it. Some conservation has taken place, but consumption has still continued to rise while production has stagnated, requiring imports of foreign oil to increase to 50 percent of total oil consumption.

The question arises: Why are we failing to achieve growth in domestic energy production?

Is it because domestic energy resources are "running out"? Hardly. The United States has potentially vast energy resources which are capable of fueling continuing economic advancement, an ever-rising standard of living and growing international strength and leadership for

15

decades—centuries—to come. The future for energy development has, in fact, never been brighter. This is why people continue to doubt that there is an energy crisis. Do we lack the technology to develop these resources? No. The United States uniquely among nations has the technical manpower and technological expertise to develop its energy resources.

But are we able to use this technology to develop our resources at reasonable cost? Evidence indicates that today's energy development will result in higher energy costs in the immediate future. However, these higher energy costs will still be affordable in relation to the value of energy to our society and the costs of other goods. Furthermore, it is entirely possible that as new, more productive resources and new, more efficient technologies are developed the cost of energy will resume its historic downward trend to where it is the same if not lower than in the past in terms of real money.

What about energy companies? Have they been holding back supplies, artificially creating shortages in order to obtain higher prices? This is a common view held by many people. However, numerous studies have been conducted by government agencies and independent investigators and *none* has corroborated this view. Energy companies may be as greedy as much of the public believes. But the people who run the companies are no more dummies than anybody else. The industry is as competitive as, if not more competitive than, most big industries, and if there is a profit to be made, somebody will produce the energy to make it.

Some people look at this picture of stagnation in domestic energy production and assume that political factionalism and governmental ineptness are at the bottom of it. And, certainly, there are difficulties in this regard. There are politics-as-usual, clashes between energy-procuring and energy-consuming regions of the country, conflicts between federal, state, county, and even city governments, and an ever-growing bureaucratic maze in which the left hand apparently does not even know the right hand exists, let alone what it is doing. Yet it is also the genius of the American political system that such problems and differences are somehow resolved and we ultimately muddle through when the views of the American public become clear, as they most certainly are today. However, in the case of domestic energy production, we are not even muddling through; we are stalled and have been for ten years.

Which brings up an important point, namely, that it is a mistake to assume that the reason energy production is going nowhere is that either nobody knows what's going on or nobody is minding the shop. Every law, every regulation, every control is today gone over with a fine-tooth comb by all kinds of pressure, interest, and other groups in today's governmental environment. Lengthy debates are oftentimes held on whether *the* or *a* should be used in a statute. With this kind of attention to detail, it may be assumed that somebody is minding the shop, that somebody does know what's going on and is planning it that way.

Every law, every regulation, and every control usually helps somebody while hurting somebody else. If this were not the case, there would be no need for the law, the regulation, or the control. Insofar as government has instituted laws, regulations and controls which have retarded energy development, it is hurting the American people by denying them needed and vital energy resources. But who is planning it this way? Who is helped by this zero energy growth?

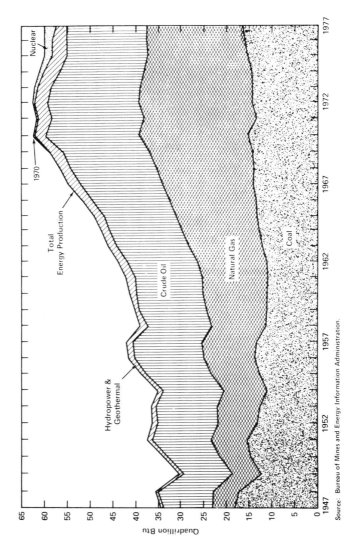

Figure 1-1. Energy Production by Primary Energy Type

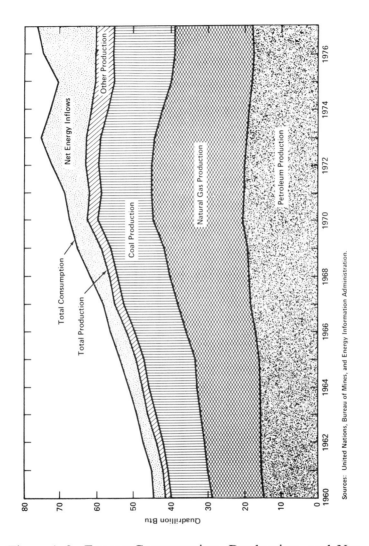

Figure 1-2. Energy Consumption, Production, and Net Inflows

To answer this question, it is necessary to examine the energy situation more closely. It was in 1970 that our total energy production peaked and went into decline. This is shown graphically in figure 1-1 in quadrillions of BTUs (British Thermal Units), the most common measure of energy. All three of our major fossil-fuel sources—oil, natural gas and coal—peaked in 1970, following which they went into a decline from which they still have not recovered as a whole. Consequently, we are faced with the startling fact that the country is producing less of its own energy today than it did in 1970, even though energy from nuclear power has been increasing.[10] Domestic energy consumption continued to increase during this period only because of increasing quantities of net energy inflows of imported oil and liquefied natural gas (figure 1-2).

What happened in 1970 to make the bottom fall out of domestic energy production? To ask the question is to answer it. The year 1970 officially ushered in the "Age of the Environment."

2

THE POWER GRAB

The Age of the Environment officially dawned in the year 1970. Environmental concern had been sparked initially by radioactive fallout from nuclear weapons testing in the atmosphere in the fifties. The late Rachel Carson's imaginative accounts of entire towns and wildlife dying wholesale of pesticide poisoning in her book *Silent Spring* (1962) created new waves of anxiety regarding environmental pollution. In the wake of this growing apprehension came federal legislation such as the Clean Air Act of 1963, the Motor Vehicle Air Pollution Control Act passed in 1965, and the Air Quality Act of 1967.

But it was not until 1970 that the environmental movement got into full swing. President Nixon's first official act of that year was taken on New Year's Day, January 1, the first day of the first year of the seventies.

While everyone else in the country was watching college football games on television, Nixon approved the National Environmental Policy Act, proclaiming that "the 1970's absolutely must be the years when America pays its debt to the past by reclaiming the purity of the air, its water and our living environment. It is literally now or never."[1]

It was a time of great excitement—and even hysteria—which peaked on Earth Day, April 22, 1970. To protect the environment from pollution, there followed the formation of a flock of new government agencies such as the Council on Environmental Quality (CEQ) and the Environmental Protection Agency (EPA) and a veritable flood of environmental legislation, including the Clean Air Act of 1970, the Clean Water Act, the Clean Air Act Amendments, the Clean Water Act Amendments, the Federal Water Pollution Control Act Amendments, the Surface Mining Control and Reclamation Act, and the Critical and Endangered Species Act.

The environmental movement also spawned an increasing concern for safety, and this resulted in the creation of government agencies such as the Occupational Safety and Health Administration (OSHA) and legislative regulations imposed by the Mine Safety and Health Act of 1970 and the Mine Health and Safety Act Amendments. All of these regulations were designed to protect workers from accidents and conditions injurious to their health.

The early seventies were also a time of budding consumerism, and the consumerists were quick to make the goals of the environmental movement their own. Ralph Nader, for example, greeted Earth Day by calling industry the worst polluter, to be curbed only by a "radical militant ethic" among consumers.[2] Inflation was also getting out of hand at the time, so, under the pressure of the consumer movement, federal price controls were instituted in 1971 to protect consumers from rising prices. All of these laws, regulations, and controls affected many areas of the country's life, but their impact on energy was particularly acute because energy production and use are historically major sources of environmental pollution. Energy production is also historically a risky, unsafe business. In addition, the cost of energy affects everyone not only because it is used by everyone but also because it is a part of the price of every product and service we buy.

The impact of these laws, regulations, and controls was to suppress or ignore the need for energy production and development. John C. Whitaker, who in 1970 led Nixon's environmental task force in the production of a presidential message on the environment, relates how in preparation of the section on air pollution what was *"completely missed* . . . was how proposed air pollution legislation would ultimately conflict with energy requirements. The task force report made only vague references to the problem, but with an emphasis on pollution control rather than potential energy shortages [emphasis added]. . . . "[3]

Passage of the Mine Safety and Health Act in 1970 along with new Environmental Protection Agency regulations resulted in the closing of 22 percent of all coal mines during 1970–71 and in a 10 percent reduction in coal output.[4]

Government price controls to "protect" consumers from rising energy prices had been applied to natural gas in 1954, resulting in continuing predictions of eventual shortages. In 1973, when controls on the prices of all other goods were removed, controls were retained not only on natural gas but oil, assuring future shortages. As Milton Friedman observes, economists "may not know much. But we do know one thing very well: how to produce shortages. . . . Do you want to produce a shortage of any product? Simply have government fix and enforce a legal *maximum* price on the product which is less than the price that would otherwise prevail. . . ."[5]

Environmental laws, safety regulations, and price controls all had the effect of putting a lid on domestic energy production, resulting in shortages and a continuing energy crisis. This is why Edward J. Mitchell, professor of business economics at the University of Michigan and director of the American Enterprise Institute's National Energy Project, says that the "energy crisis is a crisis of energy policy. It is the consequence of shortage policies adopted without reference to the public interest."[6]

Environmental laws were used to retard coal leasing, restrict surface mining of coal, and require expensive air pollution equipment when coal is used to fire power plants, prompting one observer to comment that there's "only one thing wrong with coal: you can't mine it and you can't burn it." Environmental laws also provided the basis for slowing down oil leasing programs in both onshore and offshore sites and delaying if not preventing the construction of oil refineries. Nuclear energy also came under the gun of environmental legislation to prevent so-called thermal pollution.

Safety regulations played an important role in holding back growth in coal production, but the major impact of these regulations and unending litigation related to them was to delay and even eliminate the development of nuclear power plants. Nuclear power was accused of being unsafe for a number of reasons, including the possibilities of radiation emissions during normal plant operations, cooling system failures which would release large amounts of radiation and radiation damage from nuclear wastes.

Federal price controls applied to oil and natural gas restricted output by making it uneconomic to develop new resources and expand production. No price controls were applied to coal and nuclear energy per se, but the primary use of coal and nuclear energy is to produce electricity, and the price of electricity is controlled by state and local public utility commissions. Here, consumer groups have been most vocal and active in opposing rate increases while denying electric utilities the right to charge for current construction costs of new generating plants. Both of these actions have had the effect of reducing the capability of utilities to provide for growth in electrical demand.

In addition to effectively turning off energy production, these environmental laws, safety regulations, and price controls boosted the cost of energy to consumers. For example, it costs money, lots of money, to reduce pollution from facilities such as power plants and oil refineries. It also runs up costs enormously when the construction of new energy facilities is delayed because of environmental and safety challenges. In addition, price controls have the ultimate effect of increasing rather than reducing costs because they result in shortages or the need to import much higher-priced foreign fuels or both, while causing inefficient and costly dislocations in production operations.

But the most important outcome resulting from environmental laws, safety regulations, and price controls was the transfer of increasing amounts of power over energy production and development from independent energy producers to government agencies. Environmental, safety, and price control programs have all been politically advanced as good and noble causes. However, they have also resulted in the accumulation of huge, new gobs of power by the bureaucracy. In politics, as in baseball, it is always best to keep your eye on the ball, which, in the case of politics, is power. Power is, in fact, the name of the game.

This increasing power might have been used wisely to achieve a healthy balance with the need for energy growth. Improvements in environmental protection and public and occupational safety are always desirable, all other things being equal. Price controls, although basically self-defeating, can conceivably serve at least the temporary function of cushioning the blow of rapidly rising fuel costs. But the imposition of such laws, regulations, and controls must be tempered by the need for increasing energy production.

Moderation in the use of this power was also dictated by the fact that, although the environment, safety, and prices were advanced as hot political issues, it was not as if improvements had not been made in the past without the benefit of bushelfuls of federal legislation. Speaking from forty years of experience in the environmental health field, Merril Eisenbud points out that it is "commonly thought that air pollution in all cities continued to worsen until the relatively recent intervention of the federal government. This is not true. In some cases, marked improvement occurred many years ago because of changes in fuel-burning practices. . . . New York City, which reorganized its air-pollution control activities in 1966, had already reduced its sulfur dioxide concentrations by more than 60 per cent in 1970."[7] Eisenbud points out that by 1970 the "number of occupational deaths *per unit of production* [had] been reduced by 92 per cent in a period of about 60 years."[8] Meanwhile, no reduction was made in the death rate in nuclear powerplants because no reduction was possible; there had never been a fatality in close to twenty years of nuclear power plant operations, resulting in an irreducible death rate of zero. And, up until 1971, the real price of energy had been declining for fifty years or more.

However, this increasing power over energy growth was not used wisely to strive for a balance between energy development and pollution abatement. It was used in a quixotic drive to achieve absolute environmental purity, regardless of benefits and costs and the deleterious effects on energy production and use. Coal, for example, is our most abundant energy resource. Yet, as Whitaker relates, the clean air law moved toward "expanded use of cleaner and scarcer natural gas as well as cleaner and almost four times more expensive, low-sulfur imported oil; and it moved away from our most abundant resource, coal, which at that time *could not be burned cleanly enough to meet the standards of the new law* [emphasis added]."[9]

The power was also used to attempt to achieve 100 percent safety in energy production, irrespective of the fact that no activity in life including energy production is risk-free. Nuclear power, for example, was opposed on the grounds of unsafe plant operations, even though, as Eisenbud points out, the "probability that a member of the public will be killed by a nuclear-power plant accident is roughly the same as for being killed by a meteorite, one of the rarest of all natural causes of death. If the calculations are in error by even a factor of 1,000, the risk of death by nuclear accident would still be less than that of being killed in a tornado or a hurricane!"[10]

In addition, the power was used to maintain and

27

expand price controls, even though resulting shortages of natural gas and oil in recent severe winters closed down industries, created massive unemployment and caused some people to literally freeze to death in their homes. As Mitchell comments, "The shortage of energy now facing the nation is not a problem for public policy—it is a public policy. Shortages can be eliminated very simply: remove price controls on energy."[11]

The power grab was used in an attempt to create an environmentalist paradise on earth. Imposing drastic environmental laws, safety regulations and price controls on domestic energy production basically accomplishes four things dear to the environmentalists' hearts. First, pollution resulting from existing domestic energy production facilities is reduced. Second, the possibility of pollution resulting in the future from growth in domestic energy production is eliminated by effectively throttling growth. Third, pollution resulting from energy production still required to meet growing demand is, in effect, exported to foreign countries by importing fuels necessary to meet the demand. Fourth, the increased costs caused by environmental laws, safety regulations, and price controls result in ever-increasing energy costs to consumers, reducing energy demand and the pollution that otherwise would result from increased energy use. Environmentalists thus look upon the decline in domestic energy production which they have achieved through their power grab and find it to be good.

3

THE CONSERVER
CULT

Most people view energy, if they think about it at all, as a means to an end. Energy is needed to heat and light our homes, refrigerate and cook our food, work our appliances, run our cars, buses, trucks and trains, and operate our stores, offices, factories, and farms. We freely choose our own social and economic goals and then attempt to develop the energy necessary to enable us to achieve them. The only constraints are energy accessibility and cost. Our social and economic objectives basically shape energy growth.

However, this traditional approach to energy production and development is rejected by environmentalists. They view growing energy use and resulting socio-economic advancement as a positive insult to the environment. Consequently, when the environmentalists achieved their power grab over energy production, they seized an historic opportunity to reverse the time-honored relationship between socioeconomic goals and energy. Instead of permitting social and economic objectives to determine the shape of energy development, they decided to use their power over energy to control the path of socioeconomic advancement.

Needless to say, this power to use energy to determine how other people lead their lives proved a great attraction not only to environmentalists but to a host of other groups, such as consumerist, public interest, and political activist organizations, which flocked to the environmentalist banner. However, power must always of course be used for a "desirable" public purpose in a democracy, so what was needed was a theologian/philosopher who would legitimate turning off energy production and development for the people's own good.

Enter the late E. F. Schumacher and his *Small Is Beautiful: Economics as if People Mattered*, published in 1975. A Rhodes scholar in economics, an economic advisor to the British Control Commission in postwar Germany, and, for the twenty years prior to 1971, the top economist and head of planning at the British Coal Board, Schumacher instantly became the herald of the environmentalist worldview by capturing its essence in one rather unsmall, 300-plus-page tome. This worldview basically calls for enshrining nature, "the environment," as sacred, while effectively repealing the Industrial Revolution. The first step is to eliminate the use of fossil fuel energy and nuclear power—especially nuclear power—because these energy sources not only pollute the environment but are running out anyway. What will we do without the energy provided by fossil fuels and nuclear power? Well, conserve, because we use too much energy in the first place and we don't really need all that much energy to be happy anyhow. But we will still need some kind of energy in the future, won't we? Yes, and this will be provided by developing energy resources which are part of the natural environment—the sun, the wind, crops (biomass), and animal waste which can be converted into energy. This will enable us to meet all of our energy needs that Schumacher has decided are really worth meeting. It will also make it possible to develop a new kind of society in which the high-technology, centralized energy facilities of today will be replaced by low-technology, decentralized energy sources of tomorrow, ushering in a bright new future in which man will assume his proper role as part of the environment.

Schumacher attacks the material progress created by the Industrial Revolution because of the revolutionary change it caused in the relationship between man and nature and what he believes is its apocalyptic impact on the environment and ultimately society itself. The basic problem, in his view is the "philosophical, not to say religious, changes during the last three or four centuries in man's attitude to nature. . . . Modern man does not experience himself as a part of nature but as an outside force destined to dominate and conquer it. He even talks of a battle with nature, forgetting that, if he won the battle, he would find himself on the losing side. Until quite recently, the battle seemed to go well enough to give him the illusion of unlimited powers, but not so well as to bring the possibility of total victory into view. This has now come into view, and many people, albeit only a minority, are beginning to realize what this means for the continued existence of humanity."[1]

Schumacher assails the growing consumption of fossil fuels such as coal, oil, and natural gas which fueled the Industrial Revolution because it is causing us to run out of what he calls "irreplaceable natural capital." Creating the illusion of unlimited powers nourished by astonishing scientific and technological achievements, he says, this "growing fossil fuel consumption has produced the concurrent illusion of having solved the problem of production. The latter illusion is based on the failure to distinguish between income and capital where this distinction matters most . . . namely, the irreplaceable capital which man has not made, but simply found, and without which he can do nothing."[2]

Schumacher is vehement about the need to reduce if not eliminate the consumption of fossil fuels, but he is absolutely apoplectic about the possibility of using nuclear power to provide energy in the future because of the resulting "accumulation of large amounts of highly toxic substances." No degree of prosperity could justify the accumulation of these substances which, he states, "nobody knows how to make 'safe' and which remain an incalculable danger to the whole of creation for historical or even geological ages. To do such a thing is a transgression against life itself, a transgression infinitely more serious than any crime ever perpetrated by man. The idea that a civilization could sustain itself on the basis of such a transgression is an ethical, spiritual, and metaphysical monstrosity. It means conducting the economic affairs of man as if people really did not matter at all."[3]

What does matter, Schumacher proclaims, is the "direction of research, that the direction should be towards non-violence rather than violence; towards an harmonious cooperation with nature rather than a warfare against nature; towards the noiseless, low-energy, elegant, and economical solutions normally applied in nature rather than the noisy, high-energy, brutal, wasteful, and clumsy solutions of our present-day sciences. The continuation of scientific advance in the direction of ever-increasing violence, culminating in nuclear fission and moving on to nuclear fusion, is a prospect of terror threatening the abolition of man. Yet it is not written in the stars that this must be the direction. There is also a life-giving and enhancing possibility, the conscious exploration and cultivation of all relatively non-violent, harmonious, organic methods of cooperating with that enormous, wonderful, incomprehensible system of God-given nature, of which we are a part and which we certainly have not made ourselves."[4]

To achieve this goal, according to Schumacher, it will be necessary to achieve a "profound reorientation of science and technology, which have to open their doors to wisdom and, in fact, have to incorporate wisdom into their very structure. Wisdom demands a new orientation of science and technology towards the organic, the gentle, the non-violent, the elegant and the beautiful. Peace, as has often been said is indivisible—how then could peace be built on a foundation of reckless science and violent technology?"[5]

What is it that we really require from the scientists and technologists? Schumacher answers that we need methods and equipment which are cheap enough so that they are accessible to virtually everyone, suitable for small-scale application and compatible with man's need for creativity. Apparently eliminating any form of high technology such as milling machines, airplanes, or large computers from his scheme of things with his requirement that methods and equipment be economically accessible to virtually anyone, Schumacher approvingly quotes Gandhi's statement that "I want the dumb millions of our land to be healthy and happy, and I want them to grow spiritually. . . . If we feel the need of machines, we certainly will have them. Every machine that helps every individual has a place but there should be no place for machines that concentrate power in a few hands and turn the masses into mere machine minders, if indeed they do not make them unemployed." Schumacher goes on to state that "the upper limit for the average amount of capital investment *per workplace* is probably given by the annual earnings of an able and ambitious industrial worker. That is to say, if such a man can normally earn, say, $5,000 a year, the average cost of establishing his workplace should be on no account in excess of $5,000."[6] This would eliminate most of today's industrial jobs, each of which requires an average investment of about $30,000.

Schumacher's second requirement that methods and equipment be suitable for small-scale application summarily dispenses with today's centralized mass production in favor of what he calls decentralized "production by the masses" which "mobilizes the priceless resources which are possessed by all human beings, their clever brains and skilful hands, and *supports them with first-class tools. . . .* The technology of *production by the masses*, making use of the best of modern knowledge and experience, is conducive to decentralization, compatible with the laws of ecology, gentle in its use of scarce resources, and designed to serve the human person instead of making him the servant of machines."[7]

Thus follows Schumacher's third requirement which he describes as "perhaps the most important of all" that "methods and equipment should be such as to leave ample room for human creativity." To Schumacher, this apparently means that it is okay to use tools, even "first-class tools" one supposes, but machines are *verboten*. He bases this view on what he calls "Buddhist economics" in which there are "two types of mechanisation which must be clearly distinguished: one that enhances a man's skill and power and one that turns the work of man over to a mechanical slave, leaving man in a position of having to serve the slave." How to tell one from the other? Schumacher quotes Ananda Coomaraswamy, a "man equally competent to talk about the modern West as the ancient East," to the effect that the "craftsman himself can always, if allowed to, draw the delicate distinction between the machine and the tool. The carpet loom is a tool, a contrivance for holding warp threads at a stretch for the pile to be woven round them by the craftsmen's fingers; but the power loom is a machine, and its significance as a destroyer of culture lies in the fact that it does the essentially human part of the work."[8]

But what if a worker is unskilled and therefore must use the power loom because he doesn't know how to use the carpet loom? Or what if the worker finds that in his case he is able to exercise all the "human creativity" he wishes to exercise using the power loom rather than the carpet loom? Or what if the worker is willing to sacrifice a little "human creativity" in favor of using a power loom to increase his production, income, and, ultimately, consumption? This would be wrong, according to Schumacher, because, from the standpoint of Buddhist economics, the "essence of civilization [is] not in a multiplication of wants but in the purification of human character. Character, at the same time, is formed primarily by a man's work. And work, properly conducted in conditions of human dignity and freedom, blesses those who do it and equally their products."[9] What people must do is reorient themselves to a production system based on "need" rather than "greed," Schumacher states, maintaining that "there can be 'growth' towards a limited objective, but there cannot be unlimited, generalized growth. It is more than likely, as Gandhi said, that 'Earth provides enough to satisfy every man's need, but not for every man's greed.' Permanence is incompatible with a predatory attitude which rejoices in the fact that 'what were luxuries for our fathers have become necessities for us.'"[10]

What we should seek instead, Schumacher proclaims, is "Right Livelihood," which he identifies as one of the requirements of the Buddha's Eightfold Path.[11] It is wrong, he maintains, to "measure the 'standard of living' by the amount of annual consumption, assuming all the time that a man who consumes more is 'better off' than a man who consumes less. A Buddhist economist would consider this approach excessively irrational [*sic*]: since consumption is merely a means to human well-being, the aim should be to obtain the maximum of well-being with the minimum of consumption. . . . The optimal pattern of consumption, producing a high degree of human satisfaction by means of a relatively low rate of consumption, allows people to live without great pressure and strain and to fulfill the primary injunction of Buddhist teaching: 'Cease to do evil; try to do good.'"[12]

Sums up Schumacher: "I have no doubt that it is possible to give a new direction to technological development, a direction that shall lead it back to the real needs of man, and that also means to the actual size of man. Man is small, and, therefore, small is beautiful."[13]

Taking it upon himself to define the "real needs of man" from a philosophical/theological standpoint, Schumacher established the metaphysics for a "small is beautiful" environmentalist view of the world in which energy production and use are transformed for the salvation of mankind. In this worldview, use of so-called natural capital such as fossil fuels is sharply reduced and eventually curtailed while nuclear power is immediately eliminated. In their place are put the "noiseless, low-energy, elegant and economical solutions normally applied in nature," namely, solar, wind and biomass energy, which Schumacher refers to as "renewable," or "income," energy sources in comparison to "non-renewable," or "capital," fossil fuels. Mass production using high technology is eliminated in favor of production by the masses in which goods are generated by people with skilful hands supported by "first-class tools" instead of machines. All of this, in turn, leads to a Buddhist vision of the good life in which people achieve a high degree of human satisfaction with a relatively low rate of consumption, enabling them to live without great pressure and strain while doing not evil but good.

If Schumacher provided the metaphysics for this "small is beautiful" worldview, it was up to Amory B. Lovins to produce the statistics to support it. An American who resigned a Junior Research Fellowship of Merton College, Oxford, in 1971 to become British representative of the environmentalist, U.S.-based Friends of the Earth, Inc., Lovins brought to his task impressive credentials, including his having served as consultant physicist to numerous governmental and private organizations such as the U.S. Energy Research and Development Administration, the U.S. Office of Technology Assessment, and Resources for the Future. Lovins was catapulted to fame by his presentation of the "small is beautiful" message not only in several books but also in a widely-read article called "Energy Strategy: The Road Not Taken?" published in the influential and prestigious pages of the October 1976 issue of *Foreign Affairs* magazine.

In this article, Lovins theorizes that there are only two

possible approaches to energy development. He calls these two approaches the "hard" path and the "soft" path. The hard path, which corresponds to using what Schumacher calls "non-renewable" energy sources, relies on rapid expansion of centralized high technologies to increase supplies of coal, oil, natural gas, and nuclear energy, particularly in the form of electricity. The soft path, which involves using primarily solar, wind, and biomass energy, or what Schumacher calls "renewable" fuels, combines a "prompt and serious commitment to efficient use of energy, rapid development of renewable energy sources matched in scale and in energy quality to end-use needs, and special transitional fossil-fuel technologies."[14]

Following in Schumacher's footsteps and faithfully echoing his antipathy to nuclear power, Lovins summarily rejects the hard path because "economic and sociopolitical problems lying ahead loom large, and, eventually, perhaps, insuperable."[15] The hard path also "entails serious environmental risks, many of which are poorly understood and some of which have probably not yet been thought of."[16] Approvingly quoting Friends of the Earth president David Brower that the hard path represents a policy of "strength through exhaustion,"[17] Lovins anticipates that it will result in shortages of gaseous and liquid fuels, an "increasingly inefficient fuel chain dominated by electricity generation (which wastes about two-thirds of the fuel) and coal conversion (which wastes about one-third)" and "intractable" capital costs.[18] He also warns against the use of nuclear energy because "there is no scientific basis for calculating the likelihood or the maximum long-term effects of nuclear mishaps"[19] and the "failure to stop nuclear proliferation may foreclose any energy path."[20]

In place of the hard path, Lovins first of all recommends conservation. Theoretical analysis, he says, suggests that "in the longer term, technical fixes *alone* in the United States could probably improve energy efficiency by a factor of at least three or four. A recent review of specific practical measures cogently argues that with only those technical fixes that could be implemented by about the turn of the century, we could nearly double the efficiency with which we use energy." In addition, he says, we can "make and use a smaller quantity or a different mix of the outputs themselves, thus to some degree changing (or reflecting ulterior changes in) our lifestyles . . . such 'social changes' include car-pooling, smaller cars, mass transit, bicycles, walking, opening windows, dressing to suit the weather, and extensive recycling."[21] Many analysts, he adds, now regard "modest, zero or negative growth in our rate of energy use as a realistic long-term goal."[22]

Next, Lovins urges that we follow the soft instead of the hard path. The soft path, he says, uses soft technologies which is a "textural description, intended to mean not vague, mushy, speculative or ephemeral, but rather flexible, resilient, sustainable and benign." Soft technologies "rely on renewable energy flows that are always there whether we use them or not, such as sun and wind and vegetation: on energy income, not on depletable energy capital."[23]

Lovins claims that "many genuine soft technologies are now available and economic. . . . Solar heating and, imminently, cooling head the list. . . . Secondly, exciting developments in the conversion of agricultural, forestry and urban wastes to methanol and other liquid and gaseous fuels now offer practical, economically interesting technologies sufficient to run an efficient U.S. transport sector. . . . Additional soft technologies include wind-hydraulic systems . . . which already seem likely in many design studies to compete with nuclear power in much of North America and Western Europe. . . . Recent research suggests that a largely or wholly solar economy can be constructed in the United States with straightforward soft technologies that are now demonstrated and now economic or nearly economic."[24]

To fuse into a coherent strategy the benefits of energy conservation and soft technologies, one further ingredient is needed, Lovins says. This is "transitional technologies that use fossil fuels briefly and sparingly to build a bridge to the energy-income economy of 2025. . . ." To fill this need, he recommends a "fluidized bed system for burning coal," calling it "the most exciting current development. . . . Scaled down, a fluidized bed can be a tiny household device . . . clean, strikingly simple and flexible—that can replace an ordinary furnace or grate and can recover combustion heat with an efficiency over 80 per cent."[25]

Taking an overall view of his handiwork, Lovins claims that the social structure can be "significantly shaped by the rapid deployment of soft technologies." This is because, among other reasons, these technologies "rely on renewable energy flows that are always there whether we use them or not, such as sun and wind and vegetation: on energy income, not on depletable energy capital." They also are "flexible and relatively low-technology—which does not mean unsophisticated, but rather easy to understand and use without esoteric skills, accessible rather than arcane." In addition, they are "matched in *scale* and in geographic distribution to end-use needs, taking advantage of the free distribution of most natural energy flows."[26]

Soft technology thus makes it possible to use small, local, or even domestic energy systems in place of large, centralized systems, according to Lovins. This results in "important types of economies not available to larger, more centralized systems" such as the reduction or even elimination of infrastructure costs and the virtual elimination of distribution losses.[27] In addition, soft technology provides "pluralistic consumer choice in deploying a myriad of small devices and refinements," in comparison to depending on "difficult, large-scale projects requiring a major social commitment under centralized management." For example, in the case of electricity, Lovins adds disparagingly, your "lifeline comes not from an understandable neighborhood technology run by people you know who are at your own social level, but rather from an alien, remote, and perhaps humiliatingly uncontrollable technology run by a faraway, bureaucratized, technical elite who have probably never heard of you. Decisions about who shall have how much energy at what price also become centralized—a politically dangerous trend because it divides those who use energy from those who supply and regulate it."[28]

The soft path, Lovins sums up, minimizes all fossil-fuel combustion, hedging our bets. Its "environmental impacts are relatively small, tractable and reversible. . . . The soft path distributes the technical risk among very many diverse low technologies, most of which are already known to work well. . . . Even more crucial, unilateral adoption of a soft energy path by the United States can go a long way to control nuclear proliferation—perhaps to eliminate it entirely."[29]

In conclusion, Lovins, like Schumacher, presents us with a vision of a new, more virtuous society resulting from a transformation in energy use via the soft path.

41

Underlying energy choices, he informs us, are "real but tacit choices of personal values. Those that make a high-energy society work are all too apparent. Those that could sustain life-styles of elegant frugality are not new; they are in the attic and could be dusted off and recycled. Such values as thrift, simplicity, diversity, neighborliness, humility and craftsmanship—perhaps most closely preserved in politically conservative communities—are already, as we see from the ballot box and the Census, embodied in a substantial social movement, camouflaged by its very pervasiveness. Offered the choice freely and equitably, many people would choose, as Herman Daly puts it, 'growth in things that really count rather than in things that are merely countable': choose not to transform in Duane Elgin's phrase, 'a rational concern for material well-being into an obsessive concern for unconscionable levels of material consumption.'"[30]

This Schumacher/Lovins vision of a beautiful, new world populated by human beings with low-technology, low-energy lifestyles of "elegant frugality" sustained by the sun and the wind and the earth was just what the doctor ordered for environmentalists. Friends of the Earth president David Brower, for example, enthused that "the kind of country and world a growing number of people want—and indeed, the kind we all require for sheer survival—will be less populous, more decentralized, less industrial, more agrarian. Our anxiously acquisitive consumer society will give way to a more serenely thrifty conserver society. . . . We will strike confidently and lightly along the soft solar energy path so ably scouted out by physicist Amory Lovins. Restless mobility will diminish; people will put down roots and recapture a sense of community. . . . Growthmania will yield to the realization that physical growth is wholesome only during immaturity, and that to continue beyond that point leads to malignancy or other grim devices that keep the planet from being suffocated with a surfeit. The earth will not swarm with life, but be graced with it."[31]

However, those who will no longer be "swarming" the earth with their lives in the "less populous" world of the future which Brower calls the "Conserver Society" may be forgiven for wondering whether the trip down Amory Lovins' soft path is necessary. For, to subscribe to this vision of the Conserver Society, one must believe that conventional energy sources such as oil, gas, coal, and nuclear power which currently provide 96 percent of all domestic energy production can simply be turned off without adversely affecting people's lives, jobs, and socioeconomic advancement. One must believe that continuing growth in conventional domestic energy production can be made unnecessary by massive energy conservation and reliance on so-called soft energy sources which presently provide, in effect, zero, repeat *zero*, percent of domestic energy production. One must believe that today's high-technology, centralized power systems can be economically and effectively replaced by decentralized, low-technology, neighborhood and even home systems of the future. One must further believe that, if given a choice, most people would prefer lifestyles of "elegant frugality," a la Schumacher's "Buddhist economics," versus continuing energy growth and socioeconomic advancement.

It becomes increasingly obvious that, to believe in the Conserver Society, one must, in fact, be a True Believer. This is because the goals of the Conserver Society are based not on rational evaluations of people's real needs or on objective analyses of energy capabilities but on the utopian vision of its environmentalist authors, and this utopian vision rather than socioeconomic necessities is what shapes the soft energy path of the Conserver Society. As Amory Lovins puts it, the "most important, difficult, and neglected questions of energy strategy are not mainly technical or economic but rather social and ethical. They will pose a supreme challenge to the adaptability of democratic institutions and to the vitality of our spiritual life."[32]

Social and ethical questions are normally decided by the free choices of individuals, with "energy strategy" flowing from these decisions. Lovins, however, reverses this traditional, democratic approach by designing an energy strategy which can only be used to achieve a single set of socioethical goals—those of the Conserver Society. Then, much like a used car salesman using high-pressure tactics to close a fast deal, he demands that we instantly reject the hard path while choosing his soft path, threatening us with war if not nuclear obliteration if we don't "buy now." "It is important to recognize," he states, "that the two paths are mutually exclusive. Because commitments to the first [hard path] may foreclose the second [soft path], we must soon choose one or the other—before failure to stop nuclear proliferation has foreclosed both. . . . I fear that if we do not soon make the choice, growing tensions between rich and poor countries may destroy the conditions that now make smooth attainment of a soft path possible."[33]

But the coercive, True Believer mentality of the Conserver Society is perhaps best illustrated by David Brower's summation that "there is public business to be done. We need to help the men and women who have chosen to undertake it. . . . The Conserver Society will encourage the Internal Revenue Service to encourage the public to participate in the public's political business."[34]

However, the beautiful, utopian Conserver Society in which the IRS is to serve as a watchdog to insure "public participation" is not a society at all, much less a free society. It has the major attributes rather of a cult, a cult which not only proposes its own socioeconomic and ethical goals for society but plans to coercively impose these goals on all of us by eliminating all sources of energy other than those of the soft path which lead to its vision of the future. Call it the "Conserver Cult."

4

FREEMAN AND THE FORD FOUNDATION

The Conserver Cult is the leading edge of a movement which has attracted not only environmentalists but consumerist organizations, political activist groups, and a host of other gurus of anti-energy-growth, anti-economic-growth, antitechnology, anticapitalism, and anti-American-lifestyle bents. Friends of the Earth is one of the most prominent environmentalist organizations, perhaps because its executive director David Brower is most active in promulgating his "turn the clock back" view that "we've got to search back to our last known safe landmark. I can't say exactly where it is, but I think it's back there about a century when, at the start of the Industrial Revolution, we began applying energy in vast amounts to tools with which we began tearing the environment apart."[1]

However, Friends of the Earth is only one of literally

thousands of environmentalist organizations which evince the Conserver Cult mentality to a greater or lesser degree. These include nationally known groups such as the Sierra Club, Environmental Defense Fund, National Resources Defense Council, Wilderness Society, National Wildlife Federation, Conservation Foundation and Worldwatch Institute, among many others.

Led by Ralph Nader, many consumerist organizations have also leapt aboard the Conserver Cult bandwagon because the power grab to turn off energy can also be used to attack the American corporate structure and replace it with Nader's vision of local consumer control of economic institutions. "No question about it," says Nader, "I think in the next decade we're going to rediscover smallness. We're going to rediscover it in technology— already there are movements around the world calling for an appropriate smaller-scale technology which is more responsive to self-control and local control. . . . If people get back to the earth, they can grow their own gardens, they can listen to the birds, they can feel the wind across their cheek, they can watch the sun come up."[2]

Political activist organizations such as the Clamshell Alliance, which has staged demonstrations and occupations on the site of the Seabrook nuclear plant being constructed in New Hampshire, have also joined the antienergy crusade. A loosely knit coalition of thirty to forty New England antinuclear groups, including organizations such as the Seacoast Antipollution League and the New England Coalition against Nuclear Pollution, the Clamshell Alliance aims its activities not merely at Seabrook or nuclear power but power over the production of conventional energy resources in general, points out analyst Milton Copulos of the Heritage Foundation. In its "Declaration of Nuclear Resistance," Copulos relates, the Clamshell Alliance states that the "supply of energy is a natural right and should in all cases be controlled by the people. Private monopoly must give way to public control." The Clamshell Alliance further demands that "American energy resources be focused entirely on developing solar, tidal, geothermal, wood and other forms of clean energy. . . ." But what the Clamshell Alliance is really at odds with, comments Copulos, is "the American lifestyle. They see it as wasteful and repressive. In its place, the Clamshell, and groups sympathetic with it, would impose what is termed a steady-state economy. This is one in which no economic growth takes place. Also, their emphasis on conservation and on the utilization of labor-intensive methods of manufacture would have major impacts on our current mode of living if implemented."[3]

Believers in the "limits to growth" popularized by a Club of Rome book of the same name of course find the Conserver Cult mentality most hospitable because of a shared agreement that we are either running out of energy resources or facing extinction from pollution or both. The cult also appeals to population control promoters such as Paul Ehrlich, author of *The Population Bomb*, who, it turns out, opposes plentiful energy as well as plentiful people, arguing that "plentiful energy can be considered less a benefit than a cost because of the very expensive mischief that it lets us do. . . ."[4]

Also finding much to admire in the cult are anti-technology advocates such as the famous Lewis Mumford, who categorically proclaims that "in historic perspective our current energy drain is only a part of the geotechnic devastation that has issued from massive material triumphs of every militarized and mechanized culture from the Bronze Age on." Invoking the direct use of solar energy by plants as the "organic key" to the energy crisis, Mumford states that "virtually every existing institution will have to be re-oriented to our essential life needs if we are even to restore the many human potentialities we have lost in the pursuit of power alone."[5] Meanwhile, anti-economic-growth perspectives paralleling pronouncements of the Conserver Cult are expressed by Herman E. Daly of Louisiana State University who plumps for a "steady-state economy" because "the disequilibrium between the human economy and the natural ecosystem, the congestion and pollution of our spatial dimension of existence, the congestion and pollution of our temporal dimension of existence with the resulting state of harried drivenness and stress—all these evils and more are symptomatic of the basic malady of growthmania."[6]

But perhaps John N. Cole, editor of a counterculture weekly newspaper called the *Maine Times*, best captures the worldview of a future "post-industrialism" envisaged by the Conserver Cult in an editorial called "The Future Has Arrived." Cole writes that

we were always a bit dubious of the timeliness of our ruminations about post-industrialism. Even we wondered if we might not be pushing too far into the future when we talked about revising an entire cultural value system. We advocated an end to the production-consumption-waste cycle, a stabilized, non-growth economic system and a way of life that requires living in harmony with nature. Instead of plundering the earth's non-renewable resources, we argued that our fundamental life-support systems should be based on renewable resources, less consumption, and an end to waste. There could be, we insisted, no continued depletion of a finite system like the earth's without eventual chaos. The Industrial Age, we claimed, had reached a point of excess—a point at which its costs to the family of man were greater than its benefits. When we talked about Maine as a prototype post-industrial community—a largely independent, self-sustaining, low-consumption community which would utilize the best of technology to harvest energy from renewable resources—we got a great many raised eyebrows and suggestions from friends that perhaps we shouldn't spend so much time on the topic. We also got some tough questions about what such a system might do to the present economic design. Would it mean an end to capitalism, an end to profits, an end to the exploitation pattern that has been unfolding ever since Eli Whitney invented interchangeable parts? Yes it would, we replied, because capitalism as we know it is based on growth; and we argue for an equilibrium state.[7]

As we have seen, the metaphysics and statistics for this Conserver Cult world view were respectively supplied by E. F. Schumacher and Amory Lovins. All that was necessary was to come up with the politics to promote it. This task was taken on by S. David Freeman, a lawyer who had come to Washington from the Tennessee Valley Authority in the early 1960s. Appointed to the Federal Power Commission by President John F. Kennedy and to the White House Office of Science and Technology by Presidents Lyndon B. Johnson and Richard M. Nixon, Freeman wrote President Nixon's energy message in the spring of 1971, following which he produced a book on energy under the sponsorship of the Twentieth Century Fund.[8] Published in 1974 and called *Energy: The New Era*, the book leaves no doubt of Freeman's environmentalist, antigrowth leanings a la Lovins and Schumacher.

Discussing the energy crisis, Freeman points out that America still has billions of barrels of oil and more billions of tons of coal in the ground. Why can't we, he asks, simply "switch back to domestic sources of fuel? Part of the reason is that we've already consumed much of the easily accessible oil; new sources are mostly off shore, where it takes longer to find and produce the oil. But a more fundamental constraint on the pace of production is a concern for protecting the environment. If the nation turned the fuel producers loose and let them charge as much as they pleased and drill where they pleased, we could have plenty of fuel in a few years. But at what cost? It would mean ravaging America to continue the joyride. The U.S. is finally setting in motion laws to protect our beaches from oil spills, our hillsides from ruinous strip-mining, and our air from poisons. To turn back the environmental-protection clock is an unacceptable option."[9]

But not to fear the resulting energy crisis, says Freeman, because it could be the "luckiest thing that ever happened to this country. . . . The energy crisis could be a turning point in determining America's pattern of growth, a time for reassessment and a shift to a slower and more rational pace of activities. We could stop racing through our lives to see how we finished in the competition and develop lifestyles that require less gasoline and provide more time for enjoying our friends and neighbors as well as the beauty of nature. But in order to build this America and this way of life we must find out what caused the current crisis and how we can buy enough time to make a smooth transition to an energy-conserving Post-Industrial society."[10]

Finding out "what caused the current crisis" requires from a political point of view identifying visible culprits who can be blamed for our energy woes. Freeman pinpoints the two major groups involved with energy: the producers of energy and the consumers of energy. Attacking the energy producers first, Freeman states that "government policy was to rely on the oil companies, and their policy was to put profits ahead of the public interest.

51

After all, the oil companies are not public utilities and they recognize no obligation to build refineries, drill more extensively for oil and gas, or spend money on developing alternative sources of energy if they believe the price is too low or environmental protection constraints are too stringent."[11] Freeman adds that the "energy crisis is not a giant conspiracy concocted in a smoke-filled hotel room by politicians and oil company executives. True, the energy companies stand to profit handsomely from the shortage and they will take advantage of the crisis atmosphere to gain higher prices, eliminate pollution control, and pursue corporate advantage where the public interest clashes with the industry's special interest. But there is no mystery why we are in deep trouble. The record of industry dominance of government policy amid public indifference is an open book. The crisis results from a failure of private, corporate energy policies originating in Houston and Dallas and New York and rubber stamped over the years by the Congress and a succession of presidents."[12] So much for energy producers who put profits ahead of the public interest and take advantage of energy shortages to run roughshod over government policies.

However, vile as they are, energy producers are not the real culprits of the energy crisis, according to Freeman. The true villians are energy consumers—you and I. The energy crisis arises, he says, from a "more fundamental problem than the behavior of the energy companies. To fashion a policy for the future we must understand the basic realities and the clash of values that are at the root of the matter. The most important reality is that America has been devouring energy as though it were as plentiful as water in a rainstorm. The savings we were able to achieve during the winter of 1973–74 are ample proof of our lavish waste of energy, encouraged by industry promotion with government support. Cheap, abundant energy was taken for granted. For the past decade, America increased its consumption of oil faster than it increased production. We filled the ever-growing gap by 'eating off the shelf' until domestic capacity was fully put to use and then by increasing imports."[13] Freeman goes on to say that "we cannot expect to remain energy gluttons, drawing rapidly increasing supplies from abroad, in a world where many nations are scrambling for the same limited supply."[14]

So there it is. Gluttonous consumers, "energy pigs," making lavish and wasteful use of energy are the real cause of the energy crisis, according to Freeman.

Of course, Freeman appears oblivious to the fact that the "energy savings" achieved during the winter of 1973–74 which provided "ample proof of our lavish waste of energy" were accompanied by a major recession and high levels of unemployment. He also fails to reconcile the seeming contradiction between his depiction of energy companies as putting "profits ahead of the public interest" and the availability to consumers of "cheap, abundant energy" which could be "taken for granted." If energy companies were putting "profits ahead of the public interest," whence came this "cheap, abundant energy" which could be "taken for granted?" Or, to put it another way, doesn't "cheap, abundant energy" fall within Freeman's definition of the "public interest"? But perhaps Freeman's biggest "blind spot" is represented by his failure to ascribe energy shortages to environmental laws and other governmental controls. He is well aware of their throttling effect on domestic energy production. But they are a *fait accompli*, a "given," which cannot be changed from his point of view.

Freeman, as a matter of fact, approvingly relates how

environmental constraints on the production of energy place "stringent limits on the rate at which energy fuels can be extracted from the earth no matter how high the price may go. And for good reason." He uses the risk of oil spills in offshore drilling as an example of this "good reason," adding that "what is true of off-shore drilling for petroleum is even more true of coal production where environmental constraints on growth have brought expansion to a virtual halt." Still another constraint, and a severe one, he says, is that "most of the remaining fuels are located in the public domain and are thus owned by government. . . . In the off-shore areas government has behaved much like a sensible monopolist. It has sold the fuel it owns at a pace slow enough to attract large bids from industry for what is offered. The pace pleases the environmentalists and the government budget-makers, who are a powerful combination."[15]

Freeman also maintains on grounds of safety that "despite the energy crisis we should make haste slowly in expanding atomic power production. A moratorium on new plants is advocated. It is argued that the moratorium should remain in effect until the back-up safety systems have been tested, satisfactory waste disposal plans implemented, and a secure system for safeguarding nuclear material placed in operation."[16]

In addition, Freeman describes how, in recent years, prices of fuels have been "more or less controlled by government to combat inflation. Consumers and government alike might well suppose that profits are sufficiently high to encourage an all-out search for more supply by the energy companies. Certainly, recent price increases should spur the search for more fuel. But for many years industry has taken the position that the price for natural gas, crude oil and even coal has not been high enough to warrant major expansion of the search for new sources or the construction of new mines."[17]

According to Freeman, all of these environmental, safety and price constraints find their expression in the world of "politics." The rate of growth in energy production, he says, is "in conflict with values strongly held by major segments of our society."[18] The slower growth movement, he maintains, is "perhaps the fastest growing popular movement in America."[19] Consumers, he states, tend to believe that "the energy companies are making enough profit and don't need large price increases to spur more rapid exploration. The environmentalists tend to feel that limits on production are essential to prevent destruction of the ecology. In a democratic political system, these values are bound to make themselves felt. This has been especially true in recent years as the consumer movement gained strength and the general trend toward participatory democracy gained wide acceptance." Consequently, Freeman sums up, no matter "how intensely the nation wants to increase energy production, it is not prepared to surrender its anti-inflation and environmental protection goals. . . . All things considered, energy production in the U.S. from existing sources will do well to show even small rates of growth in the years ahead."[20]

All of which of course is right on because energy production from existing sources has not only not grown at all but has actually shown a decline. But why then blame energy producers for the energy crisis when, as Freeman himself so clearly shows, the real culprit is environmental and other governmental controls? The answer is that Freeman is not so much concerned with increasing energy production as changing the behavior of his selected culprits, energy producers and consumers. Do energy producers place "profits for their stockholders ahead of supplying their customers" and "sit on their reserves if they believe they can make more money developing them later"? Then the U.S. government, Freeman states, must become the "fuel supplier of last resort." He proposes a "U.S. Fuels Corporation," which "would be empowered . . . to take whatever actions were required to assure that the nation's energy was adequate with minimum damage to the environment."[21]

Do energy consumers continue to insist that "more is always better"? Then energy consumers, Freeman states, must be persuaded to practice conservation because "the U.S. has led the world toward a pattern of energy consumption in which increasing per capita use has roughly paralleled the increasing material affluence and mobility of society. The time has come to separate the two curves, so to speak. The efficiency of energy utilization can be improved, and our society can continue to spread basic material comforts among the population without such rapid and wasteful growth in the consumption of energy."[22]

Freeman believes that we will still need energy growth in the immediate future—until 1985. "Domestic oil and gas can fill part of the gap that otherwise would be filled by Arab oil in the next ten years," he says. "And coal can supply whatever else is needed to meet the needs of a conservation-oriented pattern of U.S. energy growth."[23] However, in 1985-2000, the growth rate in energy consumption is "very much an open question," Freeman maintains. But his view is that "a realistic and desirable objective for the U.S. would be a full-employment, knowledge-intensive, food-and-service-oriented economy that would be fueled at a fairly stable level of energy consumption."[24]

According to Freeman, the continuing conservation measures that would be required to achieve this "stable level of energy consumption would be "part of a much broader restructuring of our economy and changing of our values, especially as these relate to material goods. The joy of buying something new would give way to the satisfaction of making something last. We could shift our preoccupations from possessions to intellectual and human concerns. We would have time to do more things with our hands, things we like to do. I do not foresee a hard or Spartan life in a slower, more frugal America. Indeed our energy requirements, though fairly stable, would be sufficient to provide decent housing, space conditioning the year round and basic conveniences such as a washing machine for everyone. But these material goods would move toward the background of our lives. Our satisfactions and growth would come more from our minds, spirits and relations with other people."[25] Ultimately, Freeman sums up, the "future of mankind requires the development of renewable sources of energy . . . . The real and perhaps ultimate step is to harness the sun directly or to control on earth the fusion reaction that occurs in the sun."[26]

Freeman's energy approach of turning down domestic energy production and turning up conservation measures while placing ultimate reliance on solar energy and other "renewable" energy resources provided the politics for the Schumacher/Lovins vision of the energy future. It then became the role of the Ford Foundation to put institutional muscle behind these energy politics. The foundation decided to consider the problem in late 1971 and assigned direction of an Energy Policy Project to Freeman. An advisory board was recruited to provide what Lewis H. Lapham, *Harper's* magazine editor, calls "an aura of national consensus." Lapham relates that Freeman "didn't particularly care who the foundation nominated to the advisory board. He was concerned only that the members of the board represented a sufficiently broad spectrum of opinion to sustain the illusion of impartiality."

Freeman ran his own show and produced a report reflecting his own views. "I had the *power*," Lapham quotes him as saying during an interview. "I had the *power*, and it didn't matter what any of them said." Lapham recalls that "Freeman reminded me of this point several times during the space of an hour, and I remember being taken aback by the violent emphasis that he placed on the word *power*. His enemies had been delivered into his hands, and he had thoroughly enjoyed the task of meting out God's vengeance."[27] God's vengeance, in this case, consisted of a report called *A Time to Choose*, which basically touched the same bases as Freeman's book, *Energy: The New Era,* except that it was now issued under the prestigious imprimatur of the Ford Foundation.

Attacking the "more is better" energy philosophy shared by energy producers and consumers while coming down hard in favor of conservation as a solution to U.S. energy problems, the report's major finding was that "it is desirable, technically feasible, and economical to reduce the rate of energy growth in the years ahead, at least to the levels of a long term average of about 2 per cent annually. . . . Such a conservation-oriented energy policy provides benefits in every major area of concern— avoiding shortages, protecting the environment, avoiding problems with other nations, and keeping real social costs as low as possible."[28]

The report denied that there is any relation between energy growth and growth in gross national product, stating that the "future rate of growth in the GNP is not tied to energy growth rates. Our research shows that with the implementation of actions to conserve energy in the years ahead, GNP could grow at essentially historical rates, while energy consumption grows at just under 2 per cent."[29]

The report also found that "it appears feasible, after 1985, to sustain growth in the economy without further increases in the annual consumption of energy. Such a *Zero Energy Growth* scenario can be implemented if needed for reasons of resource scarcity or environmental degradation, or it may occur as a result of policies that reflect changing attitudes and goals."[30]

According to Lapham, within a few months after the first meeting of the Ford Foundation energy policy project's advisory board, it "became obvious to several members that Freeman was directing the study to a predetermined conclusion. Freeman's choice of consultants, his casting of the argument along conservationist lines, his insistence on a program of social justice—all of it foretold the advent of an ideological tract. Phillip Hughes, the assistant comptroller general of the General Accounting Office and one of the two federal bureaucrats on the board, accepted the politics of the project as nothing out of the ordinary. 'Everybody knew at the beginning,' he said, 'that the point of view would be Freeman's.'" Harvey Brooks, the dean of engineering at Harvard who was another board member, also understood that 'what Freeman wanted all along was a political document.'"[31]

While the report was in preparation, Freeman seemingly embarrassed the Ford Foundation when he was quoted as making a number of political statements about energy and other matters during a speech to the Consumers Federation of America. As one foundation representative put it, it was "not the kind of speech you'd want a study director to make," explaining that "prejudicial statements compromised Dave's credibility as an objective and open-minded researcher into the energy question." The advisory board insisted that Freeman make a public apology, which he did by issuing a statement to the effect that he regretted the way in which his speech might have been interpreted.[32]

At its final meeting, Lapham relates, the "project's advisory board, finding that Freeman's bias was deeply embedded in the text of the report, required him to include a series of specific recommendations. Although originally conceived as an open-minded examination of all the factors relevant to an energy policy, the report quite clearly wasn't any such thing. Any pretence to objectivity might have been cause for embarrassment. . . . People were talking about its poor scholarship and about the chance of the foundation being embroiled in a political scandal."[33] Several months after the publication of the report, Lapham further notes, a "hierarch at the Ford Foundation disavowed the worth of the report, describing it as inept, foolish, and of little consequence."[34]

None of this, however, prevented the Ford Foundation from publishing the report, as Lapham observes, with "all the expensive ceremony that the foundation attaches to announcements of grave social significance. It was presented at press conferences convened simultaneously in New York and Washington. At least 6,000 copies of the report were given to members of Congress, the federal bureaucracy, and the press; during the Autumn of 1974, another 30,000 copies were sold in bookstores, and the foundation arranged for the Book-of-the-Month Club to offer an additional 300,000 copies of an abridged text to its civic-minded subscribers."[35]

With the publication of *A Time to Choose* by Ford Foundation Energy Policy Project, the Conserver Cult went national.

5

CONGRESS, CARTER AND CALIFORNIA

So why get excited about the Conserver Cult? There are lots of groups in the country and they all have programs which they are promoting. The Conserver Cult is one of many groups on the outside trying to get on the inside, right? Wrong! What's to get excited about the Conserver Cult is that its program of turning off conventional sources of domestic energy production and enforcing severe conservation measures in preparation for entry into the wonderful "soft energy" world of the future has for all practical purposes been adopted by the highest levels of government, unbeknownst to the vast majority of the American people and even against their express wishes.

This metamorphosis from the fringe to the center of power can be seen in the history of presidential and congressional politics related to energy, environmental and consumer issues over the decade of the seventies. President Nixon presided over the installation of the environmental movement in the legal framework of the land. However, Nixon was not unaware of the need for energy development, and following the 1973 oil embargo, his instinctive response was to propose "Project Independence," a program whose goal was to achieve energy self-sufficiency for the United States by 1980. Project Independence never got off the ground due to the regulatory maze in which energy development was increasingly being smothered. But Nixon remained aware of the collision course between energy and the environment and continued to try to achieve a balance. He had, for example, signed the Clean Air Amendments into law on the last day of 1970. But, when it appeared that these amendments might overly hinder energy development, he attempted to rectify the situation. John Whitaker relates that "Nixon, concerned that the balance had tipped too far in the direction of clean air at the expense of an adequate energy supply, instructed Treasury Secretary Simon, EPA Administrator Train, and the staffs on OMB and the Domestic Council to outline amendments to the 1970 act that would ensure the development of adequate sources of energy. On March 22, 1974, the administration sent a package of thirteen amendments to Congress. Very few of them were acted on. . . ."[1]

President Ford also attempted to get growth in energy production back on track by revising environmental laws and eliminating price controls. William E. Simon, who served as secretary of the treasury under Ford, tells how the "administration started out bravely enough. The President vetoed shortsighted strip mining legislation that would have cut back on desperately needed coal production. He pressed for expanded oil production in the frontier areas of Alaska and the Outer Continental Shelf. He sought to remove the regulatory roadblocks that were crippling the nuclear industry and to guarantee that by 1986, 300 nuclear plants would be supplying 20 percent of the nation's electricity. Finally, the President recommended the immediate lifting of all price controls from oil and natural gas since it is impossible to have increased exploration, production, and innovative technology when the government holds the price artificially below the market price."[2]

However, Ford met with congressional resistance and, as Simon observes, with "each round of resistance Ford cut down his own proposal, prolonged the price control period, until by about the third round of compromises, he was accepting a forty-month extension of price controls and other provisions virtually identical to those in a bill he had vetoed a year earlier." According to Simon, political "realists" at the White House then argued that "we'd better sign this or we'll lose the New Hampshire primary." The president heeded this advice and the country ended up with the Energy Policy and Conservation Act of 1975.[3]

Providing *more complex* price controls on *more* oil than had ever existed before, EPCA was, in Simon's words, a "disastrous energy law." He pointed out that the law contained no serious attempt to remove the throttles on energy production due to "frantically excessive environmental and pollution regulations." The law tacitly assumed that our energy production would be "strangled forever, and it concentrated heavily on conservation measures." Furthermore, the law established a series of "new controls for both the oil industry and for all energy-using industries" which is to say, as Simon puts it, for "virtually all industry in the United States. It contained a variety of emergency powers to permit the executive and congress to supervise energy production, to alter its nature and methods, to coerce industries into switching from one fuel to another, to allocate and ration. In effect, the law turned energy production, without saying so, into a national utility" and was a "major leap in the direction of a centralized, controlled economy."[4]

During the seventies, Congress thus took the lead in espousing laws, regulations, and controls promoting Conserver Cult interests. Presidents Nixon and Ford did what they could in the apparently thankless task of attempting to encourage energy production but were swept aside by political "realities." Whitaker, for example, recalls there was little debate within the Nixon administration about the wisdom of signing the Clean Air Amendments into law. "Besides," he comments, "given the nearly hysterical support for the environmental movement, a veto would have been futile: Congress would have promptly overridden it."[5]

When Jimmy Carter assumed the presidency in 1977, however, all presidential resistance to the Conserver Cult evaporated because the new president gave all the appearances of being a charter member. Before the election, this result had been signalled by environmental groups, which clearly favored Carter over Ford. The late David Comey of Citizens for a Better Environment, for instance, stated that Carter's "positions are so enlightened, it's like one of us running for the presidency." Meanwhile, Dan Swartzman, an environmental public service lawyer, observed that "on environment and energy, the candidates are 180 degrees opposite."[6] Two years later, environmental organizations vindicated these views by holding a public love feast in honor of President Carter. Thirty-four leaders of the nation's major environmental groups hailed the president for compiling an "outstanding" record on environmental issues. One of the leaders, Brock Evans of the Sierra Club, even went so far as to declare that if Carter maintained his performance and corrected a few environmental weak spots, he could go down in history as "the greatest conservation President in the history of the United States."[7]

Although there were undoubtedly many factors leading to this environmentalist acclaim, it is probable that none outweighed Carter's approach to energy as embodied in his administration's National Energy Act proposed in April 1977. Described in Carter's words as the "moral equivalent of war," the act turned out to be primarily a war on energy. It was nothing more than a carbon copy of the recommendations of the Ford Foundation report produced by S. David Freeman, who not so incidentally became the main assistant to Department of Energy Secretary James Schlesinger in the Carter administration.

Freeman had sent a copy of *A Time to Choose* to then-Governor Carter in late 1974, and he had heard, according to Lewis Lapham, that the governor "carried the volume around with him as if it was as precious to him as the writings of Reinhold Niebuhr." In summer 1976, Lapham goes on to report, Freeman "went to Georgia to help Mr. Carter with policy. He stayed to become a member of the team that wrote the energy plan. Accounts published in the newspapers suggest that the plan was brought forth in an atmosphere of secrecy and mistrust. It was written by government functionaries, by lawyers and academics, almost all of them, like James Schlesinger (President Carter's nominee as chief of the new energy agency), representatives of the technocratic interest. As the subsequent debate has made plain, the planners apparently didn't find it necessary to talk to people unlike themselves—not to politicians, not to officials in the

Departments of Transportation and the Treasury, not to oil-company executives, not to anybody who might violate the purity of their moral vision. . . . The plan endorsed, almost as an article of religious faith, the principle of energy conservation and reserved the government's right to restrain all forms of unauthorized development and to demand equal sacrifices from all social classes and interest groups."[8] This approach was wholly in accord with Lapham's impressions of Freeman, whom he recalls from his interview as a "slender and intense man, quite obviously possessed by a utopian vision of the just society, [who] cast his arguments in the same mode of self-righteous Puritanism that has come to characterize the rhetoric of President Carter. Freeman spoke in a quiet and slightly Southern voice, but his speech had an edge of harshness in it, as if he thought himself chosen to draw the lines of moral geography, if necessary with the point of a knife."[9]

President Carter introduced his National Energy Act to the American people in a White House television talk in which he wore a sweater to emphasize the need for energy conservation. A loincloth and turban might have been even more appropriate, for he declared that we "simply must balance our demand for energy with our rapidly shrinking resources," adding that the "oil and natural gas that we rely on for 75 percent of our energy are simply running out." Because we are now running out of gas and oil, he went on to say, we must prepare quickly for a change to "strict conservation and to the renewed use of coal and to permanent renewable energy sources like solar power."[10] In a speech to Congress two days later, Carter added that "as a last resort we must continue to use increasing amounts of nuclear energy."[11] The president of course made the obligatory swipe at the supposed gluttony of American consumers, stating that "each American uses the energy equivalent of 60 barrels of oil per person each year. Ours is the most wasteful nation on earth."[12] Later in the year, he also made the mandatory attack on energy companies, accusing them of the "biggest ripoff in energy history."[13]

Following closely the recommendations of the Freeman/Ford Foundation report, the National Energy Act proposed by President Carter established as a basic national goal for 1985 the "reduction of annual growth of United States energy demand to less than 2 percent" by "carrying out an effective conservation and fuel efficiency program in all sectors of energy use." This was necessary, the act stated, because it had been found that there was an "insufficient domestic supply of oil and natural gas." The act called for an "increase in annual coal production to at least 400 million tons over 1976 production." Finally, the act found that "the United States needs to develop renewable and essentially inexhaustible energy sources to ensure sustained long-term economic growth." Meanwhile, the potential of nuclear power to provide energy in the future was ignored in the national energy goals proposed by the act.[14]

If the fine hand of Amory Lovins as well as Freeman's input can also be detected in this proposed legislation, it is no accident. Lovins' October 1976 article on soft technology in *Foreign Affairs* was read into the *Congressional Record* by Sen. Gaylord Nelson (D-Wisc.) and passed on by friends within the Carter administration, according to writer Richard Rhodes. This led, Rhodes relates, to a "meeting [between] . . . President Carter [and Lovins] in October, 1977, and there's little doubt that the meeting, the article, and Lovins' late-1977 book, *Soft Energy Paths,* which Carter has read more than once, led to the

President's May 3, 1978, speech in Denver committing the United States to a goal of 25 percent solar power by A.D. 2000."[15] Commenting on his meeting with the president, Lovins recalls that "we talked about how to implement a soft energy path and how that kind of policy related to his energy bill. He made it clear that he's technically and politically excited by these ideas, and I was interested in, and impressed by, his depth of interest in technical matters. I have the impression that he's someone who could do much better if he were given better advice. The talk was for getting acquainted and exchanging views. I think we'll be getting together again."[16] Lovins also thought well of Carter's energy plan, commenting that "there are details of the plan that I would quibble with, but overall I'm encouraged by it. It does—by putting energy conservation first—reflect a complete reorientation of the American energy policy. That's a good first step."[17]

Energy Secretary James Schlesinger who accuses the American people of being "energy junkies who must be weaned,"[18] carried the Carter administration commitment to conservation and solar power another step forward with presentation of the Department of Energy's first award for exceptional public service to Denis Hayes of the environmentalist Worldwatch Institute.[19] National coordinator of 1970's Earth Day and chief coordinator of 1978's Sun Day, Hayes had announced in an essay that "more than one-half the current U.S. energy budget is waste. For the next quarter century the United States could meet all its new energy needs simply by improving the efficiency of existing uses."[20]

But, if President Carter caused a complete reorientation of American energy policy, he also created a revolution in government personnel overseeing energy, according to H. Peter Metzger, author and former *Denver Post* reporter, now with the Public Service Company of Colorado. Metzger points out that "there are many people around today who want to slow down the wealth-generating machine of our society—the economy—for reasons based no less on myth and superstition than the religious fanatics of another time. These people have been around in great numbers since the 1960's, but only last year the 'spoils system' operated to bring them to the seat of power. Like most zealots they are very different from ordinary people. True believers all, they exude party line and exclude from serious attention any person or opinion which doesn't conform with theirs. As you would expect, with such people in power, Washington is a changed place. There is a great unease there. Gone is the respect between rivals, and even the idea of 'the loyal opposition' has disappeared. Carter's promise of a populist, egalitarian, informal and open government has turned into a government of intolerant zealots, almost religious in the intensity of their beliefs."[21]

Metzger quotes Llewellyn King, publisher of *The Energy Daily,* who observes that "for those who have lived in Washington for a number of years and come to enjoy its institutions, its little ceremonies and the courtesies between rivals, it all leaves an unpleasant taste in the mouth. There is a feeling that bigotry, clothed in righteousness, has taken over and is fouling the processes of government. It came about because President Carter has introduced into public service a new kind of individual not formerly part of the Washington scene. They are the environmentalists, the consumer-advocates and others from what is loosely called the counterculture. The strength of the new men and women who dot the Carter Administration and who came out of a gaggle of activist organizations is that they feel in possession of moral legitimacy."[22]

Metzger relates that one campaign promise that Carter made good on was that he "hoped to challenge [Ralph Nader] in the future for the role of top consumer advocate in the country." And, says Metzger, that's just what happened. He points out that a

great many sub-cabinet posts have been given to former public-interest lawyers, consumerists, civil-rights workers and especially environmental advocates. Though their numbers are less than 100 in all, the jobs they hold are very powerful: fourteen key White House assistants—including the President's chief speech-writer—come out of the public-interest movement. . . . If the anti-energy activists had captured only White House posts it would have been serious enough but former anti-energy activists are now four Assistant Attorneys General in the Department of Justice and are Assistant Secretaries in the Departments of Health, Education and Welfare, Commerce, Interior, Agriculture, and Housing and Urban Development. Even more important, perhaps are the Naderites who themselves control large bureaucracies and multi-million-dollar budgets in completely or partially independent agencies. . . . [But] the environmentalists have had the biggest victories: Ranking jobs in the Environmental Protection Agency and the Department of Interior have gone to men and women who have sued in the courts and lobbied on the Hill for conservation, protection of wildlife, and clean air and water. *All three members* of the Council on Environmental Quality come from their ranks. And a half dozen of the most active critics of the Nixon-Ford policy on the exploration of the continental shelf are now Carter bureaucrats in various agencies[23]

But isn't this just another case of a new "old-boy network" coming in with a new president? No, says Metzger, explaining that

for the first time in history those in power have decided that the goose has layed enough golden eggs, and she's going to be retired. You may think that might be an exaggerated goal for a mere 100 people, even if the President himself is among that number, but consider how easy it is to stop something. Hundreds of people can put together a plan only to have it rejected or sent back for further study by a single key individual . . . and then studied, studied again and then restudied, etc. . . . . I call these people "Coercive Utopians." That they are Utopians is self-evident, but that's no crime. After all, many of us are or were, at least, ourselves Utopians. But, the difference between classic Utopians and these is that instead of convincing the public that their vision of tomorrow is so attractive that we ought to move their way by normal democratic means and convinced by their good example, they are doing it covertly and, therefore, coercively.[24]

However, although the Carter administration has adopted the Conserver Cult vision as its own, it cannot claim to be the leader in this regard. This dubious distinction must go to the State of California, always in the forefront of fashionable and chic new trends. In a 1976 public referendum, the people of California voted two-to-one *in favor* of nuclear power development. But no matter because, in late 1978, the state's Energy Commission killed plans for a big nuclear powerplant while putting in an order for experimental windmills! This reflected a new energy policy enthusiastically backed by Gov. Jerry Brown and designed to turn the state into the nation's biggest laboratory for energy conservation and use of alternative power sources. "We've been ahead of everybody," bragged an Energy Commission aide, commenting on the state's plans to rely on energy sources such as solar home-heating, windmills, and power generation from biomass.

Meanwhile, plans for conventional powerplants and gas installations have been scuttled, and not one major generating plant has been approved in California in the past five years. Reported the *Wall Street Journal,* "California energy commission members seem committed to a mix of conservation and alternative energy akin to the 'soft energy path' popularized by Amory Lovins. The soft-path idea also intrigues federal officials; Mr. Lovins has been warmly received by President Carter, and Energy Secretary James Schlesinger has said that future research and development budgets of his agency will focus on less-conventional energy sources."[25]

This, then is the Conserver Cult path down which government at federal and even state levels is heading. Shut off growth in conventional sources of energy such as oil, gas and nuclear power—especially nuclear power. Institute massive programs of energy conservation. Use coal only as absolutely necessary as a "transitional" fuel. Place all reliance for future energy supplies on so-called soft energy sources such as the sun and the wind and biomass.

There is only one thing wrong with this vision of our energy future. It won't work. It will result not in a "Buddhist," small-is-beautiful paradise but in an energy catastrophe, for it is based on utopian fantasies rather than energy facts.

6

ENERGY FACTS

Energy is the capability to do work. The increasing use of a growing amount of energy to do more and more work has fueled socioeconomic advancement throughout history.

Human energy is the most valuable energy resource of all because it has the capability to do creative work. This creative capability of human energy has been used to develop all other sources of energy which we use today. W. Phillips Gramm, professor of economics at Texas A&M University, points out that "to those who walked naked in the forest, the only mineral resource was a sharp stone. By using resources that were to such a man valueless, we were able to walk on the moon. Resources are created by man in the same sense as man creates anything, i.e., by rearranging and utilizing more efficiently what he finds in nature. Man creates his own environment and his own resources. . . ."[1]

However, although human energy is the most valuable energy resource on earth, it is also the most limited. It is most obviously limited by the number of people living in the world at any given time. It is even more significantly limited by the fact that a human being is a small, inefficient, and costly producer of energy. Take, for example, a man doing hard physical labor digging a ditch or plowing a field. How much power is he producing? Enough to operate a 100-hp motor? A 10-hp motor? A 1-hp motor? Far from it. The power this hard-working man produces is at best only about enough to power a 100-watt bulb, a single 100-watt bulb. This is not a big deal in the world of power.

How efficiently does this hard-working man produce energy? If he works at full speed ten hours a day, he will produce 100 watts for ten hours, which is the energy equivalent of 1,000 watts for one hour, or one kilowatt-hour. However, to produce this daily energy output of one kilowatt-hour, he requires a food-energy input of about 3,000 calories, which is the equivalent of more than three kilowatt-hours. The human body uses up two-thirds of its food energy input simply to operate its metabolic system while producing waste heat and other waste products. So more than three kilowatt-hours of food energy input are required to produce one kilowatt-hour of human energy output. This means that the man produces energy with an efficiency of about 33 percent.

But this is not the whole story. Today it takes about ten units of other kinds of energy consumed by the tractors, trucks, trains, fertilizer plants, and irrigation pumps involved in the food production chain to produce each unit of food-energy input.[2] Consequently, total energy input of ten times three-plus kilowatt-hours, more than thirty kilowatt-hours, is required to produce one kilowatt-hour of human energy output. This is an efficiency of only a little more than 3 percent, nothing to write home about.

But what does it cost for this hard working man to produce his kilowatt-hour of energy working ten hours a day? Assuming a minimum wage in the neighborhood of $3 an hour, it costs $3 an hour times 10 hours, or $30. But this is the same kilowatt-hour of energy that can be bought from an electric utility at a maximum price of about 6 cents. So if our hard-working man produces the energy, it will cost $30 divided by 6 cents—500 times more than if the energy is provided by electricity.

Producing power at a maximum rate of only 100 watts with an efficiency of only 3 percent at a cost hundreds of times greater than other energy sources, man is an extremely small, inefficient, and costly source of energy. E. F. Schumacher thinks this is a good thing, stating that "man is small, therefore, small is beautiful." However, there is nothing beautiful about this smallness to the man who needs energy to improve himself and his living conditions. Man is small, as Schumacher says. But the world is big. And man's aspirations are bigger.

It should therefore come as no surprise that the history of human advancement is based on man's successful development of bigger, more efficient, and more economical energy resources than himself. This successful development enabled him to do more and more work with less and less human energy. Even more important, it enabled him to conserve more and more of his human energy for doing the creative work of which he alone is capable. As Roger Revelle, Harvard professor of population policy, puts it: "An old saying has it, 'slavery will persist until the loom weaves itself.' All ancient civilizations, no matter how enlightened or creative, rested on slavery and grinding human labor, because human and animal muscle power were the principal forms of energy available for mechanical work. The discovery of ways to use less expensive sources of energy than human muscles made it possible for men to be free."[3]

In the beginning, the only energy resource immediately available to man aside from his own muscles was the sun. Solar radiation not only warmed his environment but gave sustenance to plants, providing the food energy to fuel his human energy. From then on, he used his human energy to develop other energy resources. Wood was the first energy resource developed by man. He burned it to obtain heat to warm his abode and cook his food. Animals provided his first source of external power. He harnessed them to pull his wagons and plows and carry him on their backs. He also learned to use windmills and sails to capture the energy of the wind, and water wheels to derive energy from falling and flowing water. But, as Dr. Richard J. Gonzalez of Stanford University points out, progress for a long time was "quite slow. So long as man was limited to primitive forms of energy, he couldn't really improve life very much. Those primitive forms were heat from the sun to grow trees and crops, wind-power to drive ships and to operate windmills for shallow wells. He also employed some falling water, his own muscle power, and draft animals. We still find that situation in primitive societies, and their way of life is not much different from what it was 2,000 years ago or 4,000 years ago."[4]

This energy picture did not begin to change until about the sixteenth century when coal came into use as a major energy source in England. John U. Nef tells how the "earliest coal-burning economy the world has known was established first in England and then in Scotland between about 1550 and 1700."[5] Previously, England had operated a wood-burning economy and experts of the time saw no reason why this should not always be the case. In *Pirotechnia* published in 1540, Vannocio Birunguccio stated that "very great forests are found everywhere, which makes one think that the ages of man would never consume them . . . especially since Nature, so very liberal, produces new ones every day." Coal was mentioned in Birunguccio's treatise but only to dismiss it as "black stones, that occur in many places, [which] have the nature of true charcoal, [but] the abundance of trees makes [it] unnecessary—to think of that faraway fuel."[6]

However, this was before the world's first major energy crisis, a wood, or deforestation, crisis, hit England. This crisis was caused by ballooning requirements for wood and wood products resulting from continuing expansion in agriculture, industry and commerce stimulated by a growing, shifting and urbanizing population. During the reigns of Elizabeth I (1558–1603) and James I (1604–25), Nef relates, pressure on the "supply of trees was reflected in the soaring cost of firewood and lumber for construction. . . . Complaints of deforestation came from all parts of the kingdom."[7] So, as the price of wood rose, the English increasingly turned to coal. Coal production and use grew rapidly and was "so successfully incorporated into the British technology and economy that during the last four decades of the 17th century wood prices stopped rising."[8]

It was England's introduction of coal as an energy resource in the seventeenth century which eventually led to the Industrial Revolution. The revolution quickly spread to the colonies which were to form the United States of America. However, up until the second half of the nineteenth century, energy sources in the United States primarily consisted of the muscle power of humans and animals along with wind, falling water, and wood.

Ninety percent of the fuel burned in 1850 was wood. Coal accounted for only about 10 percent. But the United States was then hit by a wood crisis of its own. Extensive cutting of forests in the east "raised the price of wood and increased the distance that it had to be transported to growing cities. So the demand for coal skyrocketed, until in 1885 coal surpassed wood as the dominant fuel."[9]

Another energy crisis of sorts led to the development of oil as an energy resource in the United States. Mitchell tells how illumination in America in 1800 was provided "mainly by oil lamps and candles. The fuel for the lamps was whale oil or sperm oil, both of which come from whales. The gradual exhaustion [of whales] in the first half of the 19th century caused prices to soar so that sperm oil rose from 43¢ a gallon to $2.55 a gallon in 1866, and whale oil prices had a very similar course."[10] People responded to these increasing prices by switching to cheaper substitutes, including camphene made from vegetable oils, lard oil, coal gas, and kerosene made from coal. By the 1850s, kerosene made from coal dominated the illumination market due to considerable cost advantages. But it went into a rapid decline with the discovery in 1859 of a new source for kerosene—crude oil. The price of crude oil started out at $18 to $20 a barrel, but declined with increasing production to 10 cents a barrel by the end of 1861. Kerosene made from crude quickly took over the illumination market. Meanwhile, the price of whale oil fell because of a drop in demand and ended up lower than when the "whale oil crisis" had begun in the 1800s.[11]

Crude oil of course grew tremendously in consumption with the invention of the internal combustion engine and the rapidly increasing use of automobile, truck, and rail transportation throughout the United States. Not only was crude oil economical, it was easily transported and more convenient to use than coal. Natural gas, which is oftentimes discovered at the same time as crude oil, came into increasing use in the early 1920s because it was in many instances even more economical than crude oil. In addition, it can be "transported by pipeline cheaply, the furnaces burning it are even simpler than oil-burners, and it burns with very little residue."[12]

Thomas Edison introduced the electric light bulb in 1879 and built America's first central electric power station in 1882. Electricity was also developed and eventually accepted for reasons of convenience and economy. As an ad which General Electric used for forty years expressed it: "What a convenience electric light is . . . and how much less it costs! In the days of Governor Bradford, light was so expensive that the frugal Puritan family extinguished its single candle during prayers. The early settlers had to learn to make candles themselves— the most arduous of tasks. *Your* light comes at a finger touch—and it is more than 100 times cheaper than candle light. 1¢'s worth of electricity will give much more light than a $1's worth of candles."

Coal was used as the energy source for the first central electric powerplant. In 1894, the first major hydropower project using the energy of falling water to spin a turbine and generate electricity was constructed at Niagara Falls. Based on economic considerations, oil and gas also came into use in electricity generation in plants around the country. In addition, nuclear power was developed in the forties and fifties and introduced commercially in the sixties and seventies as a way of generating electricity on a more economical, efficient basis.

Stretching over thousands of years, the history of energy development is one of continuing growth. This growth is measured by Andrew L. Simon in terms of the daily consumption by individuals of kilocalories of energy. Kilocalories are the same as the calories used to measure food energy and are equivalent to 1/860 of a kilowatt-hour. Before primitive man used fire and hunted animals, he consumed energy only in the form of plant food, and his consumption consisted of only about 2,000 kilocalories a day. This more than doubled to 5,000 kilocalories a day when he began to hunt and use fire as a source of energy. Early agricultural man more than doubled energy consumption again to 12,000 kilocalories a day. Energy use more than doubled again to 27,000 kilocalories a day when agricultural man advanced through the use of better tools and domestic animals. Early industrial man almost tripled his daily energy consumption to 70,000 kilocalories a day around 1870, while the American technological man of 1970 had more than tripled his energy use to 230,000 kilocalories a day.[13]

Because of the tremendous increase in energy use which has occurred over the years, annual energy consumption in the United States is today measured in larger units—quadrillions of BTUs. Equal to about one-quarter kilocalory, a BTU is the amount of heat required to raise the temperature of one pound of water one degree Fahrenheit. A quadrillion BTUs, or quad, is 1,000,000,000,000,000 BTUs. The quad is equivalent in energy to 175 million barrels of oil, one trillion cubic feet of natural gas, 40 million tons of bituminous coal, 2,500 tons of uranium, or 100 billion kilowatt-hours.

Energy consumption in the United States increased from 2.3 quads in 1850 to 76 quads in 1977. For the period from 1850 to 1920, increases in energy consumption per decade ranged from 25 to 73 percent and averaged 38 percent, the equivalent of an annual increase in energy consumption over the seventy-year period of about 3.3 percent. From 1920 to 1930, the decadal increase in energy consumption was only 9 percent, or less than 0.9 percent a year, reflecting the onset of the Great Depression in 1930. Growth in energy consumption per decade decreased further to 7 percent, or less than 0.7 percent a year, during the thirties, as the Depression raged in full force. In the decade of the forties, however, energy consumption resumed its historic growth, going up 41 percent, or an average of about 3.5 percent a year. In the fifties, the average annual increase in energy consumption was 3.3 percent. Energy consumption grew even faster in the sixties, at an average annual rate of 4.2 percent. And increases in energy consumption were at an average annual rate of 3.6 percent from 1970 to 1977, except for the two economic recession years of 1974 and 1975, when consumption actually declined 2.6 and 2.8 percent respectively.[14] Meanwhile, electricity consumption fueled by coal, oil, gas, nuclear power and hydropower grew at a rate approximately double that of energy in general and accounted for 19 of the 76 quads consumed in 1976.

What these continuing increases in energy consumption have meant in terms of greater human productivity and conservation of human energy is explained by Gonzalez, who points out that

the energy we use is the equivalent of having 400 mechanical servants working for us 5 days a week, 50 weeks a year. It's a total of roughly 100,000 servant-equivalent days per year per person in the United States. This is why we are productive. This is why we can have the kind of standard of living that we enjoy . . . we have one-tenth of one percent of our work being done by people, and ninety-nine and nine-tenths percent of our work being done by these inanimate servants, who are available any time we call upon them to do all kinds of things that human servants could not do for us. No number of human servants could ever transport us through the air at 500 miles an hour, or even in an automobile at 60 miles per hour. Nor could they refrigerate our food, or turn on the television set, or provide us with central heating and air conditioning. Unquestionably, these inanimate servants are terribly important in terms of our standard of living.[15]

As Gonzalez' observations suggest, growth in energy consumption over the years has been closely correlated with growth in gross national product (GNP). In good times, energy consumption grows at a rate of 3 or 4 or more percent a year. In bad times, energy growth declines to zero or below. Figure 6-1 shows how changes in energy and GNP have closely paralleled each other for the years 1947–74. The relationship between energy consumption and income per capita is even closer. Gonzalez points out that "in the United States we use three times as much energy per capita as the average for Western Europe. We use almost four times the average in Russia. We use about thirty times the average in the developing countries. As we plot the relationship of income and energy per capita for various countries, we find a very close relationship. That doesn't mean that you can increase your income just by using energy, but it means that energy is one of the requisites that you have to have in order to increase the production of goods and services. Put another way, a gallon of oil (or its equivalent in terms of coal, or gas, or hydroelectric power, or nuclear power) provides the power for the economic activities associated with two dollars of GNP, measured in terms of 1970 purchasing power."[16]

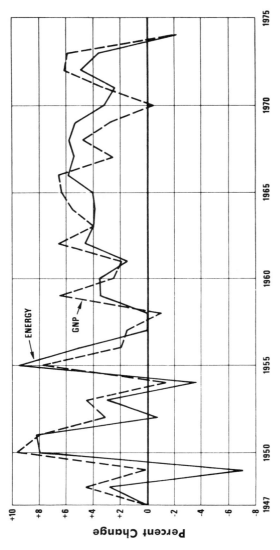

Figure 6–1. Changes in Energy and GNP, 1947–1974

Energy consumption has increased in parallel with GNP in the past even though continuing conservation efforts have successfully been made to improve the technical efficiency of energy use. Henry R. Linden, president of the Institute of Gas Technology, points out that

true conservation can only be achieved if a way can be found to lower the amount of energy required per unit of GNP. Since the quantitative interdependence of energy consumption and the economy has not yielded to U.S. and world economic changes, technological progress, changes in living patterns and habits and, particularly, to government actions in either wars or peacetime, I think it is safe to say that we do not yet know how to go about uncoupling energy consumption from GNP.[17]

The history of energy growth shows that it takes considerable time for any new energy resource to have a major impact on consumption (figure 6-2). Back in 1850, for example, fuel wood accounted for 90 percent of all energy consumption while coal provided the remaining 10 percent. Over the next sixty years, fuel wood consumption declined while coal use increased, with the result that coal accounted for almost 80 percent and fuel wood only 10 percent of energy consumption by 1910. The balance of energy consumption was made up by oil, natural gas and hydropower, together representing a little more than 10 percent. From 1910 to the present, coal has declined to near 20 percent of energy use while oil has increased to more than 40 percent and natural gas to 30 percent. The balance of consumption is currently accounted for by nuclear, hydro- and geothermal power. This historical evidence indicates that it takes fifty or more years for a new energy source to have a major impact on energy

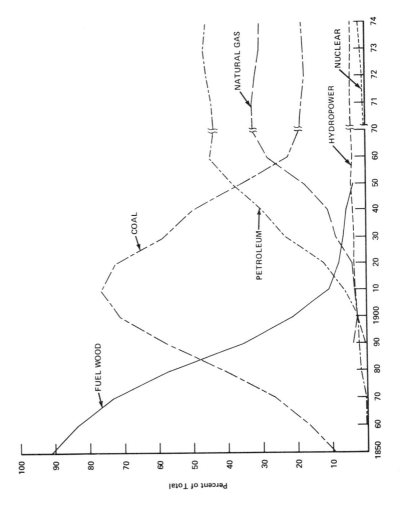

Figure 6-2. U.S. Energy Consumption Patterns, 1850–1974

consumption. This is because it takes time and money not only to develop increasing quantities of a new resource but to convert existing or develop new technology for its use.

The total growth in consumption of different energy resources is shown in quads in figure 6-3. The figure illustrates that we have not switched to new energy resources in the past because previously used resources "ran out." What has happened instead is that the price of energy resources has risen long before they were in danger of "running out," making it economical to develop other resources to take their place. Thus, consumption of coal began rising when wood was at its peak in 1850, consumption of oil and gas began increasing when coal was at its top in 1910, and consumption of nuclear power is now beginning to rise with oil and gas at current high points.

Total U.S. energy consumption during 1977 was seventy-six quads, broken down as shown in the table below.[18]

Energy Source	Units	Quads	Percent of Total
Crude oil	6.7 billion bbls.	37.00	49
Natural gas	19.2 trillion cf	19.60	26
Coal	625.0 million sht. tons	14.10	19
Nuclear	251.0 billion kW-hrs.	2.70	3
Hydro	230.0 billion kW-hrs.	2.40	3
Geothermal	3.6 billion kW-hrs.	0.08	negligible

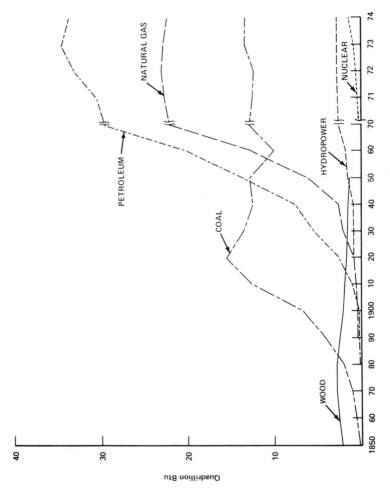

Figure 6-3. U.S. Energy Consumption Trends, 1850–1974

Total U.S. energy production consisted of sixty quads as shown in the next table.[19]

Energy Source	Units	Quads	Percent of Total
Crude oil	3.0 billion bbls.	17.40	29
Natural gas	19.9 trillion cf	21.80	36
Coal	695.0 million sht. tons	15.90	27
Nuclear	251.0 billion kW-hrs.	2.70	4
Hydro	220.0 billion kW-hrs.	2.30	4
Geothermal	3.6 billion kW-hrs.	0.08	negligible

The difference between consumption of seventy-six quads and production of sixty quads in 1977 was almost entirely due to a shortfall of crude oil production. This was more than made up by net imports of eighteen quads of crude oil. Net imports of about one quad of natural gas were also made during 1977.[20]

In the future, our energy needs will vary widely depending on our rate of energy growth. Assuming that we are using about eighty quads of energy a year as the decade of the eighties opens, the annual amounts of energy we will be using and the cumulative amounts we will have used under varying energy growth rates by the year 2000 are shown in the table below.

Annual Rate of Energy Growth	Annual Energy Use in Quads by 2000	Cumulative Energy Use in Quads by 2000
4%	175	2,477
3	144	2,214
2	119	1,983
1	98	1,779
2% to 1985; 0.0% thereafter	90	1,776

If we consume energy at the rate of 4, 3, 2, or 1 percent a year from 1980 to 2000, we will respectively use cumulative totals of 2,477, 2,214, 1,983 or 1,779 quads. The Conserver Cult, however, denies that there is any close relationship between energy consumption, GNP, and per capita income, claiming that historic growth rates of 3 to 4 percent a year in energy consumption can be eliminated by massive conservation programs while creating a better life for all. The cult therefore would increase energy use at a rate of no more than 2 percent a year until 1985 and then put a lid on growth, resulting in a total cumulative consumption of only 1,776 quads of energy by 2000. At the same time, the cult would phase out current energy sources such as oil, gas, coal, and nuclear power because they are either "running out" or environmentally damaging or unsafe, and replace them right now with "renewable" and "benign" soft energy sources such as solar, wind, and biomass. But what can be done in the real world of energy depends on what energy resources are available to us, whether we have the technology to use them and how much they will cost.

7

IS THE ENERGY THERE?

Although human energy is extremely limited in quantity, it is the energy resource with the highest quality because it is capable of doing the creative and intellectual work necessary to develop all other energy resources. Sociologist Ben Wattenberg expresses this as well as anyone when he asks us to "consider, then, the root idea of that substance we call 'natural resources.' Three hundred years ago, coal was regarded as a rock, not a resource. A hundred years ago, oil was black sticky stuff that fouled the streams of Pennsylvania (Sierra Club, where were you when we needed you?). Forty years ago, uranium ore had no known commercial value. And today, the sun shines, the tides rise and fall, the wind blows, and men in laboratories wonder how to harness these ancient natural forces that are suddenly seen as 'resources.' . . . What do all these resources have in

91

common? Simply this: They are made from something apparently useless into something quite clearly useful by the intellect of man. That is the only real resource there is. Oil in Alaska, or in Saudi Arabia for that matter, is not a resource until attacked and subdued by the intellect of man. To say that we are 'running out of resources' means that the one key resource—the intellect of man—is running dry. But that is not happening. . . . The energy-making process is more dependent on the wits of man than on the amount of gunk in the ground, and the wits of man have never quite been so titillated as they are now."[1]

There is nothing new about the doomsday thesis that we are running out of energy resources. Mitchell, for example, drily comments that "if the age of an idea contributes to its validity then the doomsday thesis has a lot going for it. However, the doomsayers have not only been consistently vocal, they have also been consistently wrong. America has had less than a dozen years' supply of oil left for a hundred years. In 1866 the United States Revenue Commission was concerned about having synthetics available when crude oil production ended; in 1891 the U.S. Geological Survey assured us there was little chance of oil in Texas; and in 1914 the Bureau of Mines estimated total future U.S. production at 6 billion barrels—we have produced that much oil every twenty months for years. Perhaps the most curious thing about these forecasts is a tendency for remaining resources to grow as we deplete existing resources. Thus, a geologist for the world's largest oil company estimated potential U.S. reserves at 110 to 165 billion barrels in 1948. In 1959, after we had consumed almost 30 billion of those barrels, he estimated 391 billion were left."[2]

One major reason why predictions continue to be made that we are going to run out of energy is a failure to distinguish between total resources and proved reserves. Total resources consist of all deposits of an energy source. Some of these deposits are known or discovered; others are unknown or not-as-yet-discovered, but their existence is inferred on the basis of reasonable scientific judgment. Proved reserves on the other hand are just a part, and a small part at that, of total resources. Proved reserves consist only of those deposits which are not only *known* but can be developed using *existing technology* at *current prices*. It takes a heavy investment in time and money to develop proved reserves. Consequently, it does not make good economic sense to develop them until they are needed, i.e., until they can be sold. Therefore, only enough reserves are maintained to assure sufficient lead times for developing new ones before existing ones run out. So, if lead times for developing new reserves are in the neighborhood of ten to fifteen years, reserves will normally amount to ten to fifteen years. And, by definition, these reserves will always be "running out" in ten to fifteen years, even though total resources may be virtually unlimited. As Mitchell comments, proved reserves are "essentially the same as what are called inventories in other businesses. The fact that oil men hold only ten or fifteen years' supply under the ground should be of as much concern to us as the fact that shoe stores keep only thirty days' supply of shoes on the shelf. To hold more would be unprofitable for the businessman and uneconomical for society."[3]

When President Carter dogmatically asserts that "we are now running out of gas and oil," he is talking about only *proved reserves,* not total resources.[4] When Energy Secretary Schlesinger comments that forecasts of enough oil and gas to last 50 to 100 years "seem to be based on smoking pot," he is basing his comment on proved reserves, *not* total resources.[5] As John Swearingen, chairman of Standard Oil Company (Indiana), puts it, when "the President and Mr. Schlesinger discuss our supplies of oil and gas, they are not talking about the total world supply, but rather about the existing *inventory* of that supply. Thus, solely in terms of inventory, many of their pessimistic predictions are not far off the mark. Given the best current figures, and given current rates of consumption and production, the existing world inventory could be used up in about 30 years, with our own domestic inventory good for about ten years. But that in no way means we are running out of oil and gas. Instead, it means that we are depleting our inventory at a fairly rapid rate. The fact is, no one knows with certainty at all how much oil and gas remain to be discovered and developed and added to our inventory in the future."[6]

President Carter relied heavily on a report prepared by the Central Intelligence Agency to back his claim that we are running out of oil. Prepared in 1977, the CIA report stated that there was only enough oil to satisfy world demand for eight more years because the Soviet Union was going to run out of oil and become a net oil importer by 1985. However, when the CIA predicted that the Soviet Union would run out of oil by 1985, it fell into what the *Chicago Tribune* calls the "'proven reserves' trap, wherein the prognosticator looks at a nation's known reserves of oil and then estimates the minimum that can be recovered *with current technology at today's prices.* But in fact higher world oil prices and advancing technology make it economic for the Soviet Union to squeeze more and more out of its reserves and to drill deeper and farther off shore for more oil."[7] The CIA later admitted its error and commissioned a new study which concluded that world oil demand would not overtake

94

supply for another sixty to ninety years.[8] The agency followed that up with yet another report claiming that mathematical equations had been formulated which confirmed that, without policy changes, the Soviets would have an oil shortage in the 1980s. But, by that time, it was questionable who, if anyone, was listening.[9]

The point is that proved reserves can be expanded first of all simply by finding new reserves which can be made available from the total resource base using existing technology at current prices. This is what has been happening to oil reserves in Saudi Arabia over the years, as is shown in the following interview with Dr. Farok Alchdar of the Saudi Central Planning Organization:

Q. Can you really be sure that you have a depletable resource? How large are your reserves?

A. In Saudi Arabia we have 140 billion barrels of oil.

Q. What was it ten years ago?

A. It depends with whom you are talking. If you were talking to the Ministry of Petroleum they would tell you about 135 or 139 billion. If you were talking to Aramco, they would tell you about 80 billion barrels.

Q. You're saying that after ten years of heavy production, you have at least as much oil as you had ten years ago by one estimate and considerably more by Aramco's estimate?

A. Well, we have discovered more resources during this time.

Q. Then how do you know that ten years from now your reserves might [not] be even greater than what they are now?

A. You are right, we might discover more oil.[10]

Other ways of expanding proved reserves include (1) coming up with new technology for developing resources previously unrecoverable with existing technology, (2) increasing prices to make it economically feasible to develop resources previously unrecoverable at current prices, and (3) a combination of both approaches.

All of these possibilities for expanding proved reserves are not only ignored but actively suppressed by the Carter administration in the apparent interest of promoting the Conserver Cult view that we are running out of energy. In June 1977, for example, soon after President Carter presented his running-out-of-energy talk to the American people, Dr. Vincent McKelvey, director of the U.S. Geological Survey, pointed out in a speech to a national energy forum in Boston that there are vast amounts of hydrocarbons sealed away in forms not presently economically recoverable. These hydrocarbons consist of natural gas located in the Rocky Mountains, in black shales in the eastern United States, and in coal beds throughout the country. He also noted that there are vast resources of natural gas in geopressurized zones of the Gulf Coast region, amounting to 60,000 to 80,000 trillion cubic feet, an "almost incomprehensibly large number." However, as columnist Paul Scott reported, it was challenges like this to the Carter administration's running-out-of-energy line which led to McKelvey's dismissal by Interior Secretary Cecil Andrus, even though McKelvey had been with the agency for thirty-seven years and had a reputation for scientific integrity. Looking into the dismissal, U.S. Rep. Jack Kemp (R-N.Y.) concluded that Dr. McKelvey had been "forced to resign his post because of his optimistic view of our energy situation—a view that was counter to the 'Malthusian' pessimism of President Carter, Secretary Andrus, and Dr. James Schlesinger, the administration's energy czar."[11]

The Carter administration's addiction to the "running

out" theory of energy was even more pronounced in the case of a Market Oriented Program Planning Study (MOPPS) prepared by the Department of Energy's Energy Research and Development Administration. Assuming correctly that increases in price would result in growing energy reserves, the MOPPS report found, in Kemp's words, that "at a price of $2.24 per million cubic feet the nation would be awash with natural gas," and "at a price of between $2.50 and $3 we would be engulfed in it." But, Kemp adds, because "this conclusion ran counter to the official line, as laid down by James Schlesinger and his mentor S. David Freeman, the Energy Research and Development Administration quickly rushed out to do another study to contradict the first. The director of the first study, Dr. Christian Knudsen, was fired."[12] But the second study also proved too optimistic, and a third study continued to show phenomenal natural gas availability. So the whole project was shucked, but not before a final Orwellian denouement reported by the *Wall Street Journal* as follows: "As we approach 1984, it will pay to watch closely the visions of George Orwell. A piece of paper crossing our desk has suddenly recalled his 'memory hole.' Information offensive to Big Brother was thrown in the memory hole, where it vanished so that history could be more conveniently rewritten. . . . We now find the following note in a distribution of the Superintendent of Documents:

ATTENTION
DEPOSITORY LIBRARIANS:

The Department of Energy has advised this office that the publication *Market Oriented Program Planning Study (MOPPS), Integrated Summary*

97

Vol. 1, Final Report, December 1977, should be removed from your shelves and *destroyed* [emphasis added]. We are advised the document contains erroneous information and is being revised. Your assistance is appreciated."[13]

Proved resources consist only of those resources which are known and can be obtained using existing technology at current prices. They do not include unknown resources which can also be obtained using existing technology at current prices. Nor do they include known or unknown resources which can be obtained as a result of improved technology and/or higher prices. Proved reserves are thus only a known and small part of total resources which are oftentimes unknowable at any given time because they are dependent on new discoveries the future may bring.

This raises a paradoxical point. So-called hard energy resources such as oil, gas, coal, and the uranium used to produce nuclear power are considered "finite" because there is only so much of them in the earth. But this finiteness is in theory only. Nobody knows just how much of each of these energy resources there really is because it does not pay to look for them until they are really needed. Consequently, we can all agree that total resources of each of these energy sources are theoretically going to run out. But nobody knows when; hell may freeze over first. By contrast, so-called soft energy resources such as solar are variously described as "renewable," "inexhaustible," "eternal," and "infinite." And it's certainly true that these energy resources are about as infinite as you can get. The sun, for example, is expected to shine every day for the next 5 billion years. But these resources are also finite, i.e., there is only so much and no more solar energy delivered to the earth each day, and solar energy "runs out" every night when the sun goes down. Soft energy resources therefore are also limited, but in a different way. And the same soft energy advocates who speak of these energy sources in glowing terms of "renewability" do not deny these limitations but rather consider them good. Amory Lovins, for example, observes that one of the "unique implications" of solar power is that "by limiting the density and the absolute amount of power at man's disposal, it would also limit the amount of ecological mischief he could do. . . ."[14] So no matter what energy resource we consider, it is limited in one way or another.

But what about currently used, so-called finite energy resources? Are they running out? Current domestic *proved* reserves of crude oil, for example, are in the neighborhood of 30 billion barrels. At present production rates of 3 billion barrels a year, these reserves will "run out" in about ten years. However, this means nothing unless we know the extent of total resources of crude oil in the country. Estimates of total domestic resources of crude oil vary widely. A panel of the Committee on Mineral Resources and the Environment of the National Academy of Sciences, for example, came up with a figure of 133 billion barrels, or 771 quads, enough to produce oil at current rates for more than forty-four years.[15] Another estimate by Herman Kahn's Hudson Institute placed total domestic oil resources at 500 billion barrels, or 2,900 quads, more than enough to provide for all U.S. energy needs at a 4 percent annual growth rate if necessary until 2000.[16]

These total oil resources include oil still remaining in existing wells which can be obtained by secondary and tertiary recovery techniques following initial recovery as well as oil yet to be discovered in onshore and offshore sites. An onshore example of how development is filled with question marks is provided by an area running roughly from Arizona to Alaska called the Thrust Belt. This area had long been considered dead because, although industry prospectors had drilled a number of wells over the last sixty years, no significant oil fields were found. However, in recent years, oil exploration technology has improved, and today there are eight producing structures in places which were thought barren just a few years back. This geological area now appears to be rich in oil and gas and is far from being completely explored.

Offshore sites provide an even greater potential for development because, although the United States has one of the longest coastlines in the world, only about 5 percent has been explored for oil.[17] What can happen as a result of aggressive exploration is illustrated by the experience of Mexico. Up until a few years ago, Mexico was not counted among major or even minor world oil powers. But, today, the country claims total oil resources of 200 billion barrels—even more than Saudi Arabia—due to the discovery that nearly 7,000 square miles along its coast and an area offshore ten times as large are potentially rich in oil and natural gas.[18] The figure may eventually be 400 billion barrels after the Mexicans finish mapping and identifying all oil-bearing formations.[19]

The highest estimate for total U.S. domestic crude oil resources of 500 billion barrels is large. But it is dwarfed by total domestic resources of oil shale. Oil shale is a sedimentary rock containing hydrocarbons which yield oil when distilled. Estimates of total shale oil resources in the United States go as high as 27,000 billion barrels, or 156,600 quads. About 90 percent of the identified shale oil resources in the United States are located in the Green River Formation, a geological formation spanning the states of Colorado, Utah, and Wyoming.[20]

The U.S. Geological Survey (USGS) estimates proved reserves for natural gas at 228 trillion cubic feet (cf). At an annual production rate of about 20 trillion cf a year, these reserves would be used up in eleven years. However, USGS also estimates that total domestic natural gas resources amount to 524 to 857 trillion cf. This is a sizeable increase over reserves. But it still does not even begin to approximate the total domestic resources of natural gas. This is because there are four additional sources in the form of methane, which is the principal constituent of natural gas. (These were mentioned by McKelvey in his Boston presentation.) One source is a huge geopressurized zone of the Gulf Coast under and near Texas and Louisiana which is estimated to contain more than 100,000 trillion cf of gas.[21] Another source is coal deposits, or beds, in which methane is trapped. This source would yield 300 trillion cf of gas from known coal deposits alone. A third source occurs in "tight sands," which are layers of clay, chalk, and sandstone. The USGS estimates that there are some 600 trillion cf trapped in reservoirs in the Rocky Mountains and a similar amount elsewhere. A final huge source of gas is found in Devonian shale in the eastern and midwestern United States, where the USGS estimates approximately 494 trillion cf of gas are trapped.[22] Adding all of these sources together provides an astounding total of well over 100,000 quads of domestic natural gas, a resource which is usually described as our most critically deficient!

Accumulating estimates for both domestic oil, shale oil, and gas resources yields a figure of 200,000 to 300,000 quads of potential energy. Even if we were able to obtain only one percent of this total energy resource between now and 2000, we would be able to completely fuel cumulative energy consumption of 2,500 quads at 4 percent annual growth rate. It therefore appears somewhat premature to declare that we are running out of oil and gas. But this conclusion is reinforced by the fact that oil and gas do not have to carry the whole load. There is coal.

Nobody even attempts to deny that total coal resources in the United States are vast. It is too common knowledge. These total resources are estimated at 4,000 billion tons which translates into something in the neighborhood of 100,000 quads of potential energy.[23] Other lower-grade forms which are also available but not in comparable quantity include lignite and peat.

Nuclear power today is fueled by uranium. Total domestic resources of uranium are on the order of more than 9 million tons, the current high estimate, with proved reserves accounting for 700,000 of this total.[24] (Nuclear energy can also be produced using thorium, which is in even greater supply than uranium.) These 9 million tons are equivalent to about 3,600 quads of electrical energy if used in today's light-water fission reactors. This does not appear to be much in comparison to some of the numbers mentioned above for fossil fuels such as oil, gas, and coal. And it isn't. However, it is only part of the story. If the same uranium were used in breeder reactors, it would produce about sixty times as much energy as it would if used in light water reactors. This results in an increase in the energy provided by uranium to 216,000 quads. But even this is not the whole story. Breeder reactors are capable of making use of otherwise wasted "tailings" produced when uranium is enriched for use as light water reactor fuel. Consisting of uranium 238, these tailings amount to 82 percent of the total uranium mined. There are today 250,000 tons of already-mined and stored uranium 238 tailings which, if converted to plutonium in breeder reactors, would provide another 15,000 to 25,000 quads. The total potential energy provided by nuclear power through breeder reactors is therefore close to 250,000 quads, or 100 times cumulative energy consumption through 2000 at a 4 percent annual energy-use growth rate.

The total resources potentially provided by these current, so-called hard energy sources—oil, gas, coal, and nuclear—are thus immense. They total up to something in the neighborhood of 600,000 quads, more than enough to power the country for many centuries to come. Hydropower is also currently used and can be easily considered a hard source because of the large, centralized dam facilities required to capture its energy. But it will be treated here as a soft energy source since it is ultimately powered by solar energy and is thus "renewable."

Overall, there is no way that these current, hard energy resources are going to run out in total over the next several centuries. But, in addition, there are other hard, or "finite," energy sources either being currently used on a limited basis or now being developed for future use. These are the potential resources of geothermal and fusion energy. Now used to meet about one-tenth of one percent of total U.S. energy demand, geothermal energy makes use of the basic heat contained in the earth. Total resources of geothermal energy in the United States to a depth of 10 kilometers (or 6.25 miles) are equal to about 24 million quads.[25] Resources of geothermal energy are thus very, very large.

But the potential for fusion energy is for all purposes inexhaustible. The primary fuel for fusion is deuterium, or heavy hydrogen. In water there is one atom of deuterium for every 6,000 atoms of ordinary hydrogen. The total energy content of deuterium used as fusion fuel is about 100,000 kilowatt-hours per gram, or one quad for every 2,200 pounds. Based on the amount of deuterium in the world's oceans, fusion energy could be used to sustain present total world energy consumption for 100 billion years or provide twenty times current world energy consumption for 5 billion years![26]

In sum, these so-called finite energy resources provide virtually unlimited sources of energy. However, the Conserver Cult would turn off these sources in favor of soft, renewable energy sources powered by the sun. And, in terms of total resources, these soft sources are also most impressive. Take, for example, solar energy itself. The amount of solar energy flowing to the United States every day is huge. The annual average for all locations is 1,450 BTUs per square foot per day. There are about 100 trillion square feet in the United States.[27] Therefore, the average daily inflow of energy from the sun is 145 quadrillion BTUs, or 145 quads. Over the course of each year, this amounts to about 53,000 quads, compared to an estimated annual U.S. consumption of 175 quads in 2000 following twenty years of energy growth at a 4 percent annual rate.

Solar radiation also drives the winds. About 2 percent of all solar energy radiated to the earth is converted to wind energy in the atmosphere. The rate at which wind energy is generated over the continental United States is about fourteen times 1973 energy demand, or a little more than 1,000 quads a year.[28]

Solar energy also activates the hydrologic cycle, in which water is evaporated from the earth's surface and redeposited in the form of rain. Assuming average rainfall, the total resources of hydroelectric energy provided by falling and flowing water in the rivers of the

United States amount to 3.4 trillion kilowatt-hours a year, or about 34 quads.[29] Hydropower can also be produced by tides or waves caused by the gravitational pull of the sun and the moon. One estimate suggests that the tidal energy in the ocean could provide about half of the current energy needs of the entire world.[30]

Solar energy is also captured by means of photosynthesis in biomass—forest, crops, other plants—which can be burned directly to produce fuel or converted into other forms of energy. A pound of dry plant tissue yields about 7,500 BTUs of heat when burned directly. A ton therefore could yield about 15 million BTUs. Under intensive cultivation, forest and field crops can yield about twenty tons of biomass per acre per year. Based on this twenty-ton yield, 3.3 million acres would be required to produce one quad of energy.[31] There are 2.3 billion acres of land in the United States. If half were used for biomass production, the total resource would amount to 320 quads per year.

Another form of biomass is organic waste, which consists of portions of municipal refuse, manure, agricultural crop waste, logging and wood manufacturing residues, sewage sludge, and some categories of industrial waste. Total resources of organic wastes are estimated at more than one billion tons per year and growing. Based on an average heat content of 5,260 BTUs per pound, this amount of organic waste can provide 10.5 or more quads of energy per year.[32]

In addition, solar energy is retained in the form of heat in the oceans. This heat can be used to produce energy in a process called ocean thermal energy conversion (OTEC), which is based on the difference in water temperature between the ocean surface and the depths below. No estimates are available of the total resources of energy available from this source. What can be said is that they are huge and could annually amount to many times energy demand.[33]

Overall, soft, renewable energy also provides potential energy resources which are for all purposes inexhaustible. Assuming total resources of about 60,000 quads a year, these sources can provide 600,000 quads every ten years, 6 million quads every century.

Thus, in the cases of both hard, "finite" energy and soft, "renewable" energy, total resources are potentially inexhaustible. Consequently, the two terms—*finite* and *renewable*—tend to fudge together and become relatively meaningless when one is attempting to understand the availability of energy resources. Lovins as much as admits this when he states in relation to hard resources that a "resource inventory would itself be useless in understanding world energy problems, for *energy constraints are not mainly dictated by physical scarcity. . . .*"[34] Use of the word *finite* to describe so-called hard resources therefore is not only misleading but potentially self-fulfilling as in the case of the Carter administration's National Energy Plan. As science journalist John Maddox says, the "sources of energy are at once more diverse and potentially more plentiful than ever before."[35] But energy plentifulness or availability means nothing if we do not have the technology to develop and use these resources.

8

"HARD" ENERGY— CAN WE GET IT?

Energy resources are abundant. There is no danger of running out of energy now or in the future due solely to physical scarcity. The energy is there. But can we get it? Do we have the technological capability to develop it?

This depends first of all on finding it. Estimates for total resources of hard energy such as oil, gas, coal, and uranium, for example, are based on the best available scientific judgment, not on precise knowledge of the exact locations of these resources. Exploratory efforts are therefore required to discover new deposits before development can even begin to take place. This problem is not faced by soft energy sources. Sun, wind, biomass, and water are obvious and, in the case of biomass in the form of animal waste, perhaps more obvious than anyone would care to see. This is probably one reason why it seems like these energy resources should be easier to develop. However, the opposite side of this coin is that, although hard energy sources are harder to find, they are easier to pack up and use somewhere else once they are found. Shipments of oil, gas, coal, and uranium are made not only throughout the country but across the oceans. By contrast, soft sources such as sun and wind energy must be used where they are found. There are no known methods for packing up and shipping sunbeams or summer breezes.

Once an energy resource is found, the next step is to develop it. The feasibility of development depends to a great extent on the specific location of the energy resource. In the case of hard energy resources, for example, technological problems vary with the depth of the resource below the surface of the earth, the geology of the area in which it is found and the type of terrain. Some coal seams may be as little as twenty feet below the surface; other deposits may be several thousand feet or more underground. The problems in drilling for oil on land are different from those involved in offshore sites. One type of technology is used to develop natural gas found under the Gulf of Mexico; another type is required for developing gas deposits trapped in mountain shale.

These effects of location on the feasibility of development may be obvious in the case of hard energy resources. But location is also a critical factor determining the feasibility of development of soft energy resources. The development potential of solar energy, for example, decreases as the distance from the equator increases. The wind blows more on the seacoast of Maine than in Cincinnati. And hydroelectric projects using the energy of falling water fare better in mountainous than in flat areas.

Another major factor determining the feasibility of development is the concentration of an energy resource.

This is the degree to which the resource occurs in any given location. Coal, for example, is found in nature in seams of varying thickness. But development is feasible only in seams which are sufficiently thick to justify the use of expensive underground or surface mining technology. Likewise, concentrations or quantities of oil and natural gas must be sufficiently great to make drilling wells worthwhile. High-grade oil shale with a concentration averaging thirty or more gallons of oil per ton of shale is feasible for development; low-grade oil shale averaging less than ten gallons per ton currently is not.

It is a characteristic of hard energy resources that they occur in sufficient concentrations in many locations to make development feasible. But the opposite tends to be the case for soft energy resources. Solar energy, for example, is very diffuse rather than concentrated. It is spread out evenly over land and water surfaces rather than being concentrated in any particular area. Wind energy is also generally diffuse, although there is a certain amount of concentration; some areas are more windy than others. Biomass is also a diffuse energy resource because it requires large land areas to collect the diffuse solar energy necessary for the growth of trees and plants. Of all the so-called soft energy resources, only hydropower is concentrated. But, in this instance, nature does the job of concentrating what otherwise is another diffuse soft energy source.

Astronomer Sir Fred Hoyle explains how "through rain and snowfalls water comes to be deposited on high ground. The rain, and the snow when it melts, then runs downhill, turning energy of position into energy of motion. At first the water runs in a multitude of tiny trickles. Then the trickles aggregate into small streams, the small streams aggregate into larger ones, until eventually a fast-moving torrent is formed. And many torrents join at last into a formidable cascading river. This aggregation of water forms the natural collecting system for hydropower."[1] The importance of this collection, or concentration, system in relation to the feasibility of hydropower development is highlighted by Hoyle when he adds that "we ourselves are not required to perform the collection. Natural drainage channels in the land surface do the collection for us, and this is precisely why hydropower is successful. If man had been required to do the sculpting of the land himself hydropower would not have been successful, for the task of shaping the land would have been too energy-consuming; and even with sufficient energy, forming the mountains themselves would surely have been beyond our capacity."[2]

The feasibility of energy development is also dependent on the quality of an energy resource. Quality is the amount of energy that an energy resource contains per unit of weight or volume or area. Hard resources provide high-quality energy because they are concentrated in energy content as well as in nature. The historical trend has, in fact, been to progress to hard energy sources of ever-higher quality. A pound of coal, for example, provides 12,000 to 14,000 BTUs when burned. A pound of oil provides 16,000 BTUs. A pound of uranium consumed in a nuclear fission reactor provides 251 million BTUs while a pound of heavy hydrogen (deuterium) converted to helium in the fusion process would theoretically yield 260,000 million BTUs.

By comparison, "soft" energy sources are characterized by low quality. Burning a pound of biomass in the form of wood, for example, yields only 5,800 to 7,700 BTUs while letting a pound of water drop over falls 1,000 feet high to produce hydropower releases a miniscule 1.3 BTUs.[3]

Another factor determining the feasibility of energy development is the efficiency with which a resource can be recovered or captured. In a typical underground coal mine, for example, about half of the total coal can initially be recovered. But the initial recovery rate is much higher for coal deposits which are surface mined. By comparison, a windmill is capable of capturing only about one-third of the energy of the wind passing through it. And the process of photosynthesis which trees and plants use to capture solar energy has a maximum efficiency of only about 3 percent and an average efficiency of only about one percent; only one percent of the solar energy shining on trees and plants is effectively used for growth.

One final energy characteristic which is critical to development feasibility is the constancy of an energy resource. Hard energy resources provide a constancy which is lacking in soft energy sources. Once a ton of coal or a barrel of oil or a cubic foot of natural gas or a pound of uranium is developed, it is available night and day, rain or shine, to provide a continuing source of energy. But soft energy is capable of serving only as an intermittent rather than constant source of energy. The sun provides its energy during the day but not at night. Wind energy is available when the wind is blowing but shuts off when the wind stops. Biomass energy is produced in the growing season but not during the rest of the year. This intermittent characteristic of soft energy sources requires that a method of storage be developed so that the energy can be used when the sun is not shining or wind not blowing or crops not growing.

The first energy resources man used were diffuse, intermittent and low-quality soft sources such as solar, wind, and biomass energy. These were all that were available to him other than his own human energy. The sun provided a warm, habitable climate and the solar energy to grow plants which could be used to fuel his own energy and the animal power of his beasts of burden. Biomass or wood was burned to provide heat to cook his food and enable him to live in colder climates. Animal wastes were also used as a fuel source just as cow dung is used today in India and other developing countries. Windmills provided the energy for pumping water from underground wells and other useful work such as grinding grain. In addition, wind provided the energy for sailing ships. For thousands of years, these were the only sources of energy used by man.

Then came the Industrial Revolution, and these soft energy sources were gradually phased out in favor of hard energy sources. Why? For all of the reasons just mentioned. "Hard" energy resources are more difficult to find, but once discovered, they can be more flexibly used to provide higher-quality, more concentrated, and more constant energy than soft energy resources. As a result, 96 percent of our energy is today derived from the hard energy resources of oil, natural gas, coal and nuclear. The remaining 4 percent of energy production comes from the soft energy source of hydropower. But this is a case where nature has already collaborated in the work of concentrating this resource, leaving only the job of capturing its energy to man. However, hydropower is as inflexible as other soft energy resources since it must be used where it is concentrated. This is why it is primarily found in mountainous areas such as the Pacific Northwest.

The technology for using hard energy resources has been developed and improved over a century or two. Typically, this technological development proceeds through a number of time-consuming stages. First, the scientific feasibility of using the technology is established. Then, engineering development is required to establish the practicality of technology use. Next, an engineering demonstration is conducted in which the technology is used as planned under actual conditions. Up to this point of development, the technology has not been used to produce any energy in the real world. Yet the period over which this technological development takes place can be ten, twenty, thirty, or more years. Following the engineering demonstration, commercial introduction of the technology takes place. The technology is offered for sale in competition with other forms of energy technology. If people buy and use the technology, and if the new technology is superior in performance, cost, or some other respect, growth in use of the technology takes place. But this growth will usually be relatively slow because people must normally throw out their old technology before they install the new technology. It usually does not make good economic sense to throw out bought-and-paid-for old technology until its useful life is over, even though new technology may be superior; so another ten, twenty, thirty, or more years may go by before the new technology begins to make a significant impact on commercial energy production.

Over the last 200 years, technology for using hard energy resources has been developed through the above stages and is now proven by experience. In addition, technology is being developed to expand the availability of hard energy resources in the future. In the case of oil, for example, there is proven, available technology for every phase of energy development, including exploration, drilling, and recovery, and this technology is constantly being improved. Seismic exploration techniques, for instance, are undergoing rapid improvement in terms of field equipment, better and more powerful computers, and more sophisticated computer programs.[4] Satellite observations are also providing an improved tool for oil exploration, pinpointing locations that otherwise would not be explored.[5] In addition, oil wells are being drilled onshore to ever-greater depths. In 1908, for instance, the deepest onshore well drilled in the United States went down less than 5,700 feet in comparison to today's deepest of 31,000 feet.[6] Offshore oil wells are also being drilled not only to greater depths but in ever-deeper waters. Currently, offshore wells are drilled from platforms in water depths of up to 500 feet. Potentially, these platforms can be used in water depths up to 1,000 feet. Beyond this depth, technology exists for drilling wells on the ocean floor by remote control.[7]

New types of recovery techniques are also being developed to obtain more oil from existing wells. Oil is found not in pools but in porous rock from which it is forced up to the surface by natural pressure when a well is drilled. However, only an average of about 32 percent of the total oil in the well is obtained during this initial recovery. Technology for secondary recovery involving waterflooding of the well has therefore been developed to force more oil to the surface. Waterflooding normally will raise the total amount of oil recovered from a well to about 50 percent, leaving half of the oil still in the ground. A number of methods of tertiary recovery, primarily involving the use of chemicals, are now being used or tested to further increase the total quantity of oil obtained from a well. One test being conducted in Wyoming, for example, is expected to recover one-third of the remaining oil.[8]

Technology for developing shale oil is proven, although there are other production methods in the engineering development or scientific feasibility stage. More than 120 years ago during the great "whale oil crisis," there were fifty-three companies producing oil from shale in the United States. However, the shale oil was expensive, and the companies went out of business when cheap conventional oil was discovered in Pennsylvania in 1859. But shale oil has never been forgotten, and over the years some 3,000 patents relating to oil shale technology have been granted.[9] Currently, several pilot plants are in operation using strip-mining technology. The shale is mined, crushed and heated, or retorted, in a kiln to produce a thick crude oil which is upgraded and refined to produce conventional oil products.[10] Another method of development involves *in situ* (in-place) processing. In this method, a mine shaft is dug into the shale and the shale is broken up underground with chemical explosives. The rubble is then ignited, creating heat which frees shale oil for pumping to the surface. This technique is now undergoing engineering demonstration at a site in Colorado.[11] Other development methods in the engineering demonstration to scientific feasibility stages involve production of other products such as alumina (an oxide of aluminum) as well as shale oil, retorting shale oil in the presence of hydrogen gas to produce natural gas as well as shale oil,[12] and simply sticking pipes into shale deposits to convert them into giant ovens that cook the oil out of the rock.[13]

Natural gas is found in both onshore and offshore sites in association with oil and by itself. All of the exploratory and drilling technology developed for oil is also applied to the discovery and extraction of conventional natural gas. In addition, a new gas exploration method called the "bright spot" technique has proven extremely useful in identifying potentially productive gas reservoirs.[14]

Recovery from gas wells is much greater than from oil

wells, ranging from 50 to 90 percent, with one to 3.4 percent being lost through flaring, venting, and production operations. Technology is also available for extracting methane from geopressurized zones, with recovery rates ranging from 4 to 50 percent.[15] Currently being vented to the atmosphere, methane trapped in coal deposits can be easily exploited with existing technology. The developed technique of hydraulic fracturing can be used to extract natural gas from tight sands and also from Devonian shale. In the case of tight sands, it is estimated that almost one trillion cf of gas a year could be produced within seven years.[16] Readily recoverable reserves in the Devonian shale would amount to 15 to 25 trillion cf over a fifteen- to twenty-year period with half again as much produced over the next 10 to 30 years.[17]

Coal is recovered using either underground or surface mining technologies, each of which accounts for about half of total production. Underground mining is primarily done using a room-and-pillar method in which about 45 to 50 percent of coal in place is recovered. Remaining coal is used as pillars to support roofs in underground rooms where mining is done. However, the recovery percentage can possibly be increased to as much as 80 percent by removing additional coal when a mine is closed down and roof support is no longer a problem.[18] Another underground method called longwall mining which is currently used for only about 3 percent of U.S. coal production provides a recovery efficiency of 85 percent because it eliminates the need for pillars.[19] Surface, or strip, mining meanwhile results in recovery efficiencies averaging 80 to as much as 98 percent.[20]

In addition to being burned directly as a fuel, coal can also be converted to a gaseous or liquid fuel. Coal gasification processes are classified as low-BTU, intermediate-BTU or high-BTU, according to the energy content or quality of the gas produced. Low-BTU methods of coal gas production were in use a century ago, but were largely displaced by cheap natural gas. Two techniques for intermediate-BTU gasification called the Lurgi process and the Koppers-Totzek process are in commercial use today on a modest scale. High-BTU gas containing mostly methane can be produced from coal in several ways and a number of small pilot plants are in operation to test these processes. However, no commercial methane plants are in operation, and continued research is needed to improve coal gasification technology.

When coal is turned into a liquid, the resulting fuel is generally a synthetic crude oil that can be refined into conventional oil refinery products. A number of different processes have been proposed for coal liquefication and several have been tested in pilot operations in the United States. Continued research, however, is needed to improve coal liquefication technology.[21]

The newest hard energy resource, nuclear fission, uses a number of different solidly proven technologies. Uranium is discovered using exploration technology similar to that used in any mining activity with one notable exception. Because it is slightly radioactive, uranium ore can be detected by sensitive instruments from considerable distances, even, in some cases, from airplanes. Uranium ore is recovered through surface, underground, or solution mining techniques. The ore is crushed, ground, and dissolved to extract uranium oxide, which is dried to a concentrate known as "yellowcake." Only 0.7 percent of the yellowcake is fissionable uranium 235. The remainder, mostly U-238, is not fissionable. This low concentration of U-235 is not enough to sustain a fission reaction. So the yellowcake is converted to a gas and enriched to about 3 percent U-235 at one of three gaseous diffusion plants owned by the federal government. This process converts 18 percent of the gas into 3 percent enriched uranium, leaving 82 percent in the form of depleted "tailings" made up primarily of U-238, which are stored for possible future use. The enriched gas is converted to uranium dioxide powder and formed into half-inch pellets, which are placed in thin metal tubes twelve to fourteen feet long. As many as 50,000 of these fuel rods are then used to form the fuel core of a large nuclear fission reactor.[22]

Two types of light water reactors make up the majority of commercial reactors in use in the United States: the boiling water reactor (BWR) and the pressurized water reactor (PWR). These reactors provide the conditions under which the nuclear fission reaction can take place. The reaction is used to heat water and produce steam which spins a turbine. The turbine drives a generator to produce electricity.[23] About once a year, one-fourth to one-third of the fuel rods in a reactor are replaced because they have used up their capability to sustain a nuclear fission reaction efficiently. Stored in pools of water for safekeeping, these rods contain both fissionable U-235 uranium not consumed during nuclear reactions and newly generated fissionable plutonium, both of which can be recovered through reprocessing.[24] If recovered and reused in nuclear fission reactors, this uranium and plutonium can provide energy for producing up to 50 percent more electricity.[25]

Mining, enriching, and fission reactor technologies have been developed, proved, and put into commercial application over the last twenty to thirty years. The technology is also constantly being improved. For example, two new enrichment technologies using ultracentrifuge and laser instead of gaseous diffusion methods are now being developed. In addition, a high temperature gas reactor (HTGR) system that uses helium instead of water as a heat transfer medium and both thorium and uranium as fuel was commercially introduced in 1966 and is now producing electric power in two locations, with ten more plants on order.[26]

Construction has taken place on three spent-fuel reprocessing plants, but due to technical, regulatory, and funding problems, none have gone into operation.[27] Following reprocessing, there are two alternative methods for recycling uranium and plutonium. In one method, the uranium and plutonium are combined into a "mixed oxide" fuel which is used in current nuclear fission reactors. In the other method, only the uranium is used to produce fuel for current reactors while the plutonium is stored for future use in fast breeder reactors.[28]

A fast breeder reactor produces heat and ultimately electricity by fissioning plutonium mixed with uranium 238. As the plutonium fissions, it produces five pounds of additional plutonium from the U-238 for every four pounds of plutonium consumed.[29] In effect, the fast breeder reactor "gives birth" to more plutonium fuel than it consumes. Capable of making use of U-238 in enrichment tailings as well as plutonium reprocessed from used fission fuel, technology for a liquid fast metal breeder reactor is now entering the engineering demonstration stage. A demonstration plant called the Clinch River Reactor Plant is currently under construction in Tennessee. However, fast breeder reactor technology is much farther advanced in other countries such as England,

Russia, and France (where a commercial-size plant is now in operation).

Fusion, the nuclear reaction that powers the sun, has been under investigation for several decades. In the fifties, a project was established to investigate the possibility of generating energy by fusing atoms of deuterium and tritium (forms of hydrogen), both of which are relatively easy to obtain, in a controlled manner within a reactor. Present strategies for developing fusion reactors involve two concepts: magnetic confinement and laser implosion.[30] Magnetic confinement has received the most attention to-date and a major breakthrough was achieved in 1978 at Princeton University's Plasma Physics Laboratory when the temperature inside a small, experimental, eighteen-inch diameter fusion reactor was pushed to more than 60 million degrees, four times hotter than the sun's interior.[31] In 1975 fusion was achieved by KMS Industries, Inc., Ann Arbor, Michigan, using lasers.[32] However, these successes only established the scientific feasibility of fusion power which is now undergoing engineering development.

Geothermal energy, the natural heat of the earth, is available anywhere if wells are dug deep enough. However, technology for tapping geothermal energy is currently developed only for use where local geologic conditions have concentrated the diffuse heat of the earth into hot spots or thermal reservoirs relatively close to the surface. These thermal reservoirs fall into three types: hydrothermal, geopressured, and dry hot rock reservoirs. Developed technology is currently being used to produce electric power in the hydrothermal Geysers area of northern California. However, the commercially attractive dry steam produced in this area is relatively rare. Areas producing hot water with steam or wet steam such as the Old Faithful geyser in Yellowstone National Park are twenty times more common.[33] Geothermal energy is produced from hydrothermal areas by drilling wells and using the dry or wet steam which rises to the surface to drive a turbine. Wells decline in production with time, so additional wells must be continually drilled to maintain steam supply.[34] No production from geopressured reservoirs has occurred to date, although these reservoirs differ from hydrothermal reservoirs only in the source of heat and thus use the same technology.[35] Technology for obtaining geothermal energy from dry hot rock reservoirs is still in an engineering demonstration stage. In these reservoirs, a well is drilled and water is pumped into it under high pressure to hydraulically fracture the dry hot rock. A second well is then drilled and water is pumped down one well and returned up the other as steam to run a turbine.[36]

Technology for producing currently used hard energy resources is not only proven but presently supplying energy in the huge quantities required to meet 96 percent of society's needs. The technology cannot be expanded overnight to increase production of these currently used energy resources. It takes years, in some cases a decade or more, to explore for deposits, drill wells, sink mine shafts, and build plants. Nor can it be stated with certainty when technologies now under development will be available to provide new types of hard energy supplies. But the energy resources are there, and there is no way to find out what increases in energy production can be achieved except through continuing development.

However, the Conserver Cult would turn off production and development of these hard energy resources and replace them right now with soft energy resources. This is the view expressed by David Brower of Friends of the Earth in relation to nuclear power, for example, when he says that "we've shown leadership in suggesting that we just skip nuclear power altogether, and get on with developing safe, lasting energy sources—solar and wind power. . . . We're convinced that nuclear power just isn't worth it. It entails risks that no one can afford."[37] It is also the view of Amory Lovins in relation to fusion when he declares that "for three reasons we ought not to pursue fusion. First, it generally produces copious fast neutrons that can and probably would be used to make bomb materials. Second, if it turns out to be rather 'dirty,' as most fusion experts expect, we shall probably use it anyway, whereas if it is clean, we shall so overuse it that the resulting heat release will alter global climate: we should prefer energy sources that give us enough for our needs while denying us the excesses of concentrated energy with which we might do mischief to the earth or to each other. Third, fusion is a clever way to do something we don't really want to do, namely to find *yet another* complex, costly, large-scale, centralized, high-technology way to make electricity—all of which goes in the wrong direction."[38] The "right direction," of course, is towards increasing use of soft energy resources. But what is the status of technology for the development of these soft energy resources so prized by the Conserver Cult?

9

"SOFT" ENERGY—
CAN WE GET IT?

Hard energy is based on the use of relatively new resources which have been discovered in the last two or three hundred years. Coal use goes back many centuries, but it did not come into major prominence as a source of energy until about the seventeenth century. Oil and natural gas were never used until the nineteenth century. Nuclear fission power is a twentieth century phenomenon. These new hard sources of energy combined with new technology have fueled the material, social, and political advancement of the past 200 years.

By contrast, soft energy is based on resources which were used prior to the advent of hard energy. Solar, wind, and biomass energy were the only energy sources available to people for thousands of years before they were gradually phased out in favor of hard energy over the past several centuries. Perhaps this is why the call to make greater use of soft energy in the future is sometimes expressed as if it were part of a "back-to-nature" or "postindustrialism" movement. Yet, because of its diffuse, intermittent, low-quality form—the reason for its decline in the past—soft energy is as much, if not more, dependent than hard energy on the development of sophisticated, state-of-the-art, high technology if it is to become a viable source of energy in the future.

Soft energy resources consist either of direct solar energy in the form of solar radiation or indirect solar energy such as wind, biomass, and hydropower. Direct solar energy can be used in either what are called passive or active solar systems. Passive solar systems have been used for thousands of years. The technology for these systems basically consists of the buildings in which we live and work and learn and play. A building can be designed and building materials used to take maximum advantage of the sun's energy to heat the building in the winter and provide cooling and ventilation in the summer. Passive design and materials are typified by large areas of south-facing glass, massive structural elements such as thick concrete walls and floors serving as heat sinks and energy-conserving insulation techniques. This type of solar system is called "passive" because the thermal flow of energy for building heating and cooling is provided by natural convection, conduction, and radiation rather than by compressors, pumps, or fans.[1] Passive technology is available in terms of new architectural designs and existing and new types of building materials. However, new construction is required to achieve its full benefits since it is difficult to effectively modify existing buildings and structures along passive solar lines. Overall, passive solar systems are simply ways of evening out temperature variations and conserving energy rather than a method of providing for the complete heating and cooling needs.

Active solar systems use some type of combined collection, storage, and distribution facilities for water heating and space heating and cooling. The most common type of collection system consists of flat plate collectors, or panels made of aluminum and glass which are placed on a sloping roof facing the sun. Because of the diffuseness of solar energy, hundreds of square feet of these panels are required for a typical house. The panels capture heat from the sun and transfer it to water or air circulating in pipes at temperatures of about 250 degrees Fahrenheit. Because of the need to store heat during cloudy days and at night, the hot water or air in the pipes is first pumped to a storage area consisting of a large tank in the case of hot water or an even larger bin containing crushed rocks in the case of hot air. The hot water or air is then circulated as needed for water or space heating or to operate a heat pump for cooling. Active solar systems are usually designed to provide only 40 to at most 80 percent of the total comfort needs of a building because installation of collector panels and a storage system large enough to do the whole job would be too costly. The balance of heating and cooling needs must therefore come from a conventional gas or oil furnace, back-up electric heat pump or electric heater.[2] Solar water and combined water and space heating systems are now being used in many parts of the country, but a large number of these installations are for the purpose of engineering demonstration. Solar systems providing cooling are also in the engineering development and demonstration stages. Extensive research is being conducted to improve the efficiency of solar collection and storage methods used in these systems.

Solar energy can be used to produce not only heat but

electricity in what are called solar thermal electric power systems. In one approach, large numbers of flat-plate collectors are mounted on the ground and the heat they collect is transferred to water in a system of pipes to produce steam. The steam is conveyed via the pipes to a central turbine, which operates a generator to produce electricity. Because of the diffuseness of solar energy, a vast area of land must be used in this approach. A power-plant using a flat-plate collector system would require twenty-five to fifty square *miles* of collectors to produce 1,000 megawatts of electricity in the sunny southwestern United States.[3] Some type of storage method would also be required with this system as with any solar system so that electricity could be provided when the sun is not shining. Technology for a solar thermal electric power system such as this is in engineering development and demonstration stages.

To improve the efficiency of solar thermal electric power systems, another type of solar energy collector called a parabolic mirror reflector can be used. This curved mirror concentrates the rays of the sun at a single focal point, producing maximum heat intensity. In addition, the mirror can be mechanically rotated so that it follows the sun across the sky, capturing as much solar energy as possible. The heat created by this approach can raise the temperature of water in collector pipes located at the focal points of such mirrors to 600–800 degrees. Because of this greater concentration of solar energy, a solar thermal electric power system of this type would require "only" ten instead of twenty-five to fifty square miles of reflectors to produce 1,000 megawatts of electricity.[4] Technology for a solar thermal electric power system using curved-mirror reflectors is also in the engineering development and demonstration stages.

To further improve the efficiency of solar energy collection while reducing land requirements, another solar thermal electric power system using a central receiver, or "power tower," has been conceived. The basic design of this system consists of an array of mirrors distributed around a 1,000-foot-high tower. The mirrors are aimed to focus sunlight on a boiler on top of the tower. A fluid such as water is heated in the boiler by the intensely concentrated, reflected sunlight from the mirrors to produce steam for operating a turbine. Temperatures up to 1,500 degrees can be obtained using this approach, which requires "only" five to seven square miles of mirrors.[5] Technology for this type of solar-thermal electric power system is also in engineering development and demonstration stages.

These solar thermal electric power systems all rely on direct collection of solar energy. However, solar energy is also collected and to some extent concentrated naturally in the oceans of the world, and there are several present and potential ways of using this energy to produce electricity. One way is to use the tidal energy resulting from the gravitational pull of the sun and moon. A tidal power system operates like a hydroelectric dam. A dam is constructed across a coastal inlet where there is a large tidal variation. As the tide comes in and goes out, it is run through turbines in the dam to produce electric power. Two tidal power plants are currently operating in Russia and France. However, there are only a few other locations in the world (with only one location in the United States, in Alaska) where diffuse tidal energy is sufficiently concentrated in terms of the rise and fall of tides to make this technology practical.[6] Another possibility is to use the energy concentrated by the up-and-down motion of waves on the surface of the ocean to run turbines for electricity production. But this idea is not much advanced beyond the engineering development stage.

Oceans also absorb energy in the form of heat. This gives rise to the possibility of another method of electrical production called ocean thermal energy conversion (OTEC) in warm ocean areas such as the Gulf of Mexico and off the coasts of Florida and southern California where surface water temperatures are relatively high. OTEC uses the difference between these high surface temperatures and colder water temperatures below the surface to generate electricity by spinning a turbine with steam in the same way as a conventional powerplant. The steam is produced using a fluid such as ammonia or freon, which has a low boiling point causing the liquid to turn to vapor at ocean surface temperatures of 70 degrees. After the vapor has passed through and spun a turbine, it is cooled and condensed back into a liquid by cold water pumped up from the depths below, following which the cycle is repeated. Operating with an efficiency of only 2 to 4 percent, a 1,000 megawatt OTEC plant would be a massive floating structure drawing on the heat stored by the sun's rays in 50 to 100 square miles of ocean. One OTEC plant producing 22 kilowatts of electricity was built in Cuba in 1924 but could not compete economically with fossil-fuel plants of the period. Currently, OTEC plants of the size required in today's world are in the scientific feasibility stage, with much of the required technology still to be developed.[7]

All of the above solar methods of producing electricity use the sun's heat to produce steam and drive a turbine as in a conventional power plant. However, solar energy can also be converted directly into electricity using transistor-like devices called solar photovoltaic cells, or solar cells.

Primarily made of silicon and developed as a major source of electric power for space vehicles, solar cells can be made to generate electricity simply by placing them in sunlight and connecting electrical wiring. However, silicon cells are only about 11 percent efficient in transforming the sun's rays into electrical energy. In addition, as in any solar energy system, some method of storage is required to provide a constant source of energy while taking full advantage of solar cell capabilities.[8] Solar cells are out of the engineering demonstration stage, and extensive research is now being conducted to improve their efficiency and ease of manufacture. But technology for electrical storage is only in the engineering development and demonstration stage.

One possible way around the efficiency and storage problems of solar cells would be to put power plants equipped with the cells in space, where the solar energy is not only stronger but available all the time (which is one reason why the cells are successfully used in space vehicles). Dr. Peter Glaser of Arthur D. Little, Inc., for example, has suggested putting into orbit solar power satellites with 25-square-mile arrays of solar cells. These huge spacestations would generate 10,000 megawatts of electricity, which would be transmitted by microwave beam to 36-square-mile receiver antennas on the ground.[9] However, the technology for this concept is in the scientific feasibility stage, with engineering development and demonstration still to take place.

Indirect solar energy in the form of wind energy has been used to pump water, grind grain, and do other mechanical jobs for many hundreds of years. The first windmill to generate electricity was put into operation in Denmark in 1890. Harnessing the wind is quite straightforward—in principle. The wind turns a windmill or large propeller which is connected directly to an electrical generator. Windmills are capable of usefully capturing about one-third of the energy in the wind passing through them. Because of the diffuseness of wind energy, windmills have to be fairly large to be effective. A windmill with a diameter of 50 feet, for example, will generate only about one-tenth of a megawatt in a 30-mile-an-hour wind while a 200-foot-diameter windmill will generate about 1.5 megawatts.

Wind also varies by location and is variable at any given location. This variability is inconvenient because even a relatively small change in wind speed causes large, sudden fluctuations in the electrical output of a wind generator, a condition which is intolerable in an electrical supply system. Some method of storing the electrical energy produced by a windmill is therefore required not only to smooth out production but to keep surplus energy produced on windy days for use on calm days.

The diffuseness of wind energy also makes it necessary to use a great number of windmills and a large land area to achieve significant energy production. In one design, a total of 1,800 windmills with 50-foot diameters on 850-foot, 75-story, towers covering 1,800 square miles would be required to produce the 1,000 megawatts generated by the average conventional powerplant. Windmill technology has been developed and used over the years, and research is being conducted to develop it further. But the crucial technology for energy storage is still in the engineering and demonstration stages.[10]

Indirect solar energy in the form of biomass consists of both the products of organic farms planted with trees and crops and organic waste, including forest and agricultural residues, animal waste and garbage. Biomass can not only be burned to produce heat but it can also be biochemically or thermochemically converted to produce medium- and high-BTU gas, alcohol, fuel oil, and other petrochemical substitutes.[11] Biomass, of course, has been used for thousands of years and continues to be used today as a source of energy. However, the process of photosynthesis by which solar energy is captured in biomass has an efficiency of only about 3 percent at peak growing times and one percent on the average, even in the case of high-yield crops. In addition, because of the diffuseness of solar energy, vast areas of land are required to produce biomass energy. It is estimated, for example, that 650,000 acres of land managed and harvested in a manner similar to a Southern pulpwood farm would be required to fuel a 1,000-megawatt powerplant.[12] Research is aimed at using more energy-intensive trees and crops such as eucalyptus trees, sugar cane, and kenaf and also algae grown on water.[13]

Organic waste is also a diffuse energy source except when it is collected as part of an existing process. This is the case when it is collected as residue resulting from operations in the forest products industry or concentrated waste in large animal feed lots or mounds of garbage in city pick-up operations. In these cases, biomass is currently being used to produce either heat for operating turbines and producing electricity or methane gas for fuel applications. Much of the technology required for producing energy from biomass is available, although improvements in biochemical and thermochemical conversion techniques continue to be made.

Indirect solar energy in the form of hydropower is produced by damming a river or harnessing a waterfall and forcing the water to flow through turbines connected to electric generators. The ideal site for a hydroelectric plant is a deep canyon with a good water flow where a relatively small dam can back up a great quantity of water into a reservoir to attain a large head or height of water and thus extract maximum energy from the flowing water. Even though nature conspires to concentrate this diffuse source of energy as a river, hydropower is still a low-quality energy source, making it necessary to use anywhere from 1,000 to 100,000 acres of land to store water in reservoirs behind dams.[14] In addition, there is a limited number of natural sites which provide the conditions necessary for efficient hydropower development. But the technology required for hydropower development is far advanced.

Aside from hydropower, these soft energy sources are currently being used to provide next-to-nothing of the country's energy production. Furthermore, technology for producing soft energy is in many cases still in engineering development and demonstration or scientific feasibility stages. The technology cannot be developed for practical production of energy overnight. It will take decades to implement it to the point where it can have a significant impact on energy production. However, the energy is there, and there is no way to find how much of it can ultimately be made available except through continuing development efforts.

Consequently, the logical approach to overall energy development is to develop technologies for all available energy resources, both hard and soft. Each energy resource has its own potential and place. There is every reason to think that hard and soft energy technologies can complement each other. There is no reason to think they cannot coexist. In particular, it makes little sense to summarily dismiss hard energy resources from future energy development.

132

But the Conserver Cult does not take this view. The Cult would turn off production of hard energy right now and place reliance on the rapid development of soft energy resources for future energy production. This is the approach taken by Lovins, who not only wants to turn off hard energy but even eliminate large, centralized soft energy technologies because "schemes . . . such as making electricity from huge collectors in the desert, or from temperature differences in the oceans, or from Brooklyn Bridge–like satellites in outer space do not satisfy our criteria, for they are ingenious high-technology ways to supply energy in a form and at a scale inappropriate to most end-use needs."[15]

But, although the availability of soft energy resources is great, technological capabilities for developing these energy resources to enable them to significantly replace hard energy cannot be developed and implemented for several decades, much less immediately. The same Conserver Cultists who claim differently oftentimes deride nuclear power because after twenty to thirty years of development it today provides only 15 percent of the electricity in the country and only 4 percent of total energy used. Yet soft energy sources are in about the same technological position today that nuclear power was in the forties.

Claims that soft energy sources are capable of significantly replacing hard energy sources right now can only be made by overstating the technological case for their development. In his *Foreign Affairs* article, for example, Lovins suggests that if we built all new houses for the next twelve years with passive solar collectors—large south windows or glass-covered, black south walls—we would "save about as much energy as we expect to recover from the Alaskan North Slope."[16] However, energy consultant

Ralph Lapp points out that "serious analysis of Lovins' assertion that solar heating through use of 'passive collectors' such as large south windows or glass-covered, black south walls would save as much energy as we would expect to recover from the Alaskan North Slope reveals that his estimate of possible energy savings is 20 times too high."[17]

In another instance, relating to biomass, Lovins claims that a conversion industry roughly ten to fourteen times the scale of the U.S. beer and wine industry could "produce roughly one-third of the present gasoline requirements of the United States."[18] But Bruce Adkins, principal administrator of the Organization for Economic Cooperation and Development, points out that this would be true only if beer and wine consisted of "pure alcohol rather than liquids containing an average of 10% alcohol, which immediately puts up the required production by 10 times, i.e., to 140 times what it is now."[19]

Drs. Aden Meinel and Marjorie Meinel, long-time University of Arizona solar energy researchers with impeccable credentials in the field, put it even more bluntly in commenting on the overdrawn technological capabilities ascribed to soft energy by Lovins in his *Foreign Affairs* piece. "We are chilled at the actions advocated by this seductive and well-written article," they state, "and appalled that a person claiming to be a physicist could resort to what appears to us as distortions of technical reality. The danger of such an article is that the appeal to a return to the 'good old days'—ultimately to the Garden itself—plays upon a deep-rooted desire of the human species. Should this siren philosophy be heard and believed we can perceive the onset of a New Dark Age. To abandon high technology and large-scale cooperative organization is a step toward making mankind once more a slave to the dictates of a dispassionate environment and his own furies, with luxury for a few and depths of despair for the masses."[20]

Sheldon H. Butt, president of the Solar Energy Industries Association who as an industry representative would appear to have every reason to take as favorable as possible a view of the potential of solar energy, provides just as blunt and uncompromising a critique of Lovins' views concerning the technological capabilities of soft energy:

> My fundamental reaction . . . is profound shock. It is shocking to me that Lovins, who has obviously benefitted from an excellent education in the physical sciences, has allowed advocacy of his own concept of the future to lead him to support this position with a lengthy series of distortions and even misrepresentations of physical and scientific fact. This is doubly shocking because Lovins' training in the physical sciences should make him sufficiently aware of the need to consider factors objectively. . . . We feel that it is of transcendent importance that we maintain a society based upon individual election, individual initiative and individual incentive, and, therefore, we support and will continue to support an appropriate mix of both the energy technologies which Lovins identifies as "soft" and "hard." We perceive that any effort to limit the development of new energy resources to those somewhat arbitrarily classified as "soft" and, even beyond this, to abandon our existing "hard" technology assets, must inevitably require that we compel individuals to conform to Lovins' societal fantasies.[21]

Furthermore, it would also require that we greatly increase the cost of energy, the third factor which is traditionally considered along with availability and technology in evaluating the relative usefulness of energy resources.

10

HOW MUCH WILL
IT COST?

Energy resources are abundant. The technology to develop them is either available or in the process of being developed. However, what energy resources we will actually use in the future depends to a great extent on price. Normally, for obvious reasons of economy, the goal is to use energy resources that provide the most energy at the lowest cost.

Historically, the real price of energy has declined over time. This decline has not been smooth. There have always been temporary increases in price such as occurred during the "whale oil crisis" of more than 100 years ago and at other times throughout our energy history. But energy prices eventually came back down and resumed their historic decline for two reasons. First, when the price of an energy resource goes up, an incentive is created to produce more of it. The resulting greater supply not only

136

dampens the price rise but can cause the price to retreat to a point at or below its former level, depending on demand and market conditions.

Second, and more important, an increase in the price of an energy resource creates an incentive to find and develop other, cheaper sources that can take the place of the resource with the rising price. This is the real reason why energy prices have historically declined over the years. As the price of wood increased in the nineteenth century, for example, lower-cost coal increasingly took its place. So also coal was replaced by lower-cost oil and natural gas in this century, and in the case of electrical production, all three of these fossil fuels are today being substituted for to a growing extent by lower-cost nuclear power.

The long-term trend of real energy prices is one of decline, and it is entirely possible that this long-term decline will resume in the future with the development of new, less-expensive energy resources. However, we are currently experiencing an "energy crisis" in which real prices are rising rather than falling. To determine what is going on, a little background is required. From 1950 to 1973, real energy prices declined at an accelerating rate. Then, *in one year,* they shot back up to 1950 levels. As Edward J. Mitchell describes it, from "1950 to the middle of 1973, prices fell continuously, and in every five-year period fell faster than in the previous five-year period. Up to the middle of 1973, we had an accelerating decline in energy prices that left the real price 25 percent below 1950 levels. In the next year, energy prices gained back all of that decline and a bit more."[1]

Was this rapid increase in prices due to a suddenly developing physical scarcity of energy resources? Not at all. It was due entirely to the actions of OPEC, using its cartel power to increase oil prices, even as OPEC members continued to pump cheap oil from the ground at a cost of twenty-five cents a barrel. Mitchell comments that "this pattern of accelerating decline and extremely sharp reversal has little to do with the physical scarcity of oil. Prices fell in the '50s and '60s primarily because competition in the world oil market was driving crude prices down toward costs of production. Prices have risen in the past year because competition in the world oil market was brought to a halt when an organization of oil-exporting countries began to operate effectively as a cartel. The shift from competition to cartel accounts for the upsurge in world energy prices."[2]

Normally, this rapid rise in price would have spurred feverish activity to increase U.S. domestic energy production due to increased financial incentives. But the price rise occurred several years after the Conserver Cult had begun turning off domestic energy production. This left the American public holding a stick with two short ends. On the one hand, we were forced to increasingly depend for our growing energy needs on foreign producers with cartel power to raise prices practically at will. On the other hand, growth in domestic energy production—which could have broken the power of the cartel—was effectively throttled.

What levels of prices would be required to substantially increase domestic energy production? Due to inflation as well as varying market demand, energy prices are changing every day. However, it is possible to provide "ballpark" figures based on the best available estimates of increases in energy supplies that would result from rising prices. In the case of conventional oil, for example, the 1977 MOPPS (Market Oriented Program Planning Study) analysis suggested that supplies would increase approximately as follows as price rose in increments of five dollars a barrel:[3]

Price (dollars per barrel)	Cumulative Quantity (billions of barrels)
$ 5	18
10	38
15	90
20	115
25	128

In mid–1979, the price of domestic oil ranged from $5 to $13 a barrel, OPEC oil sold at $16 to $17 a barrel, and spot prices in the international market were $20 a barrel and up.

In addition, as the price of conventional oil goes up, it becomes increasingly more economical to develop shale oil resources. A range of oil shale production cost estimates based on three different production methods is shown below in 1978 dollars:[4]

Production Method	Estimated Production Cost per Barrel (1978 dollars)
Underground mining and surface retorting	$ 7.23 to $10.27
Surface mining and retorting	6.70 to 9.80
Conventional *in situ*	10.59 to 16.73

The 1977 MOPPS study also considered the effect that rising price would have on the economic availability of

139

conventional domestic natural gas. Following are estimates of the cumulative natural gas in trillions of cubic feet which would become economically available with price increases in increments of one dollar per thousand cf:[5]

Price (per thousand cf)	Cumulative Quantity (trillions of cf)
$1	120
2	275
3	475
4	580
5	625

In mid-1979, conventional domestic natural gas sold for as little as forty cents per thousand cf on long-term contract and for two dollars or more per thousand cf in uncontrolled markets. But the cumulative quantities shown above reflect estimates only of conventional domestic natural gas which would become economically available with rising price. The estimates do not include unconventional sources of domestic natural gas such as geopressurized methane, methane in coal beds, gas in tight sands, and gas in Devonian shale. A 1978 analysis predicted that production of these additional huge gas resources would become economical at prices of between two and three dollars per thousand cf.[6]

Proved reserves of coal are available in tremendous amounts that are more than sufficient to supply our needs for many years to come. Since proved resources by definition are resources available at current prices, growing supplies of coal can be obtained as long as adjustments for inflation, productivity and other variable factors are made in coal prices. In 1978, these prices ranged from about $20 a ton for medium-BTU, high-sulfur coal to approximately $35 a ton for high-BTU, low-sulfur coal.[7] In addition, coal can be used to manufacture synthetic oil and gas fuels whose production becomes more economically feasible as the prices of oil and natural gas rise. In terms of oil, for example, this would involve a rise to a price of $20 to $25 a barrel.

Economically available quantities of domestic uranium for producing nuclear electric power also increase as prices rise. The Department of Energy made the following 1978 estimates for cumulative available uranium in tons of reserves at increasing production costs in dollars per pound as shown below:[8]

Production Cost (per pound)	Cumulative Reserves and Potential Resources (tons)
$15	1,565,000
30	3,255,000
50	4,365,000

In mid-1979, uranium was selling for $40 to $45 per pound.

These four hard energy resources—oil, natural gas, coal, and uranium for producing nuclear-electric power—account for practically all of our domestic energy production. Normally, rises in prices of these resources would spur increased supplies because the increasing prices would make it economic for producers not only to expand existing production but to develop higher-cost oil and gas fields and coal and uranium deposits. In addition, rising prices would create greater financial incentives for the development of other domestic energy resources. This expanding domestic energy production would eventually reach a point where increasing supplies would begin to diminish the market for OPEC products and undercut the cartel's power to set prices. This would occur at prices greater than today's but less than if OPEC were permitted to continue to operate its cartel. Furthermore, this price rise could initiate the development of new energy resources with the long-range potential of resuming the historical decline in real energy prices.

However, this normal market reaction to rising prices has effectively been aborted by more and more stringent environmental laws, absolute-safety regulations, and comprehensive and complicated price controls. The result has been a decrease in domestic energy production and a corresponding sharp increase in energy prices as growing energy demand has had to increasingly be met with imports of more and more costly foreign fuels. Meanwhile, the Conserver Cult that engineered this expensive result simultaneously points with alarm to rising hard energy costs while beckoning us to come and partake of the "cheap" energy provided by soft sources such as the sun, wind and biomass.

"Dear Friend, Fellow Tax Payer and Earth Resident," opens David Brower in a Friends of the Earth form letter. "If you had your choice, would you rather . . . have a low-cost, locally manufactured solar unit on your roof or centrally located in your neighborhood to give you clean, environmentally sound energy to heat and light and cool your home, or . . . keep on paying your utility company 12 months a year, year in and year out, for the privilege of buying energy from high-cost, controversial nuclear plants? The former is cheaper, simpler and gives you more independence and control over the quality of your life. The second is one more way to be endlessly obligated to your utility company and its ever-increasing rates."[9]

Amory Lovins proclaims that "many genuine soft technologies are now available and are now economic . . . Solar heating and, imminently, cooling head the list. . . ."[10] Recent research suggests that a largely or wholly solar economy can be constructed in the United States with straightforward soft technologies that are now demonstrated and now economic or near economic."[11]

Other advocates enthuse that soft energy is cheaper than hard energy because it's "free." The sun and the wind and growing things are simply there. Nobody has to pay anything to get them, right? Nope. Frederick H. Morse of the University of Maryland and Melvin K. Simmons of Lawrence Berkeley Laboratory point out that "although it is often said solar energy is free, this is no more true for solar energy than for any other energy source. Oil in the ground costs us nothing: nature and time have provided it. However, to extract, transport, refine, and distribute oil to the consumer costs us in both labor and capital. The same is true of solar energy: the cost is in the labor and capital required to make it useful to our needs. Solar energy does differ from most of our present energy sources in that it is very intensive in 'first cost.'"[12]

What about first, or capital, costs of "hard" versus "soft" energy? In his *Foreign Affairs* article, Lovins envisages a hard energy path consuming about 160 quads by the year 2000. He declares that "among the most intractable barriers to implementing [this hard energy path] is its capital cost."[13] By contrast, he proposes a soft energy path consuming only 95 quads in 2000. Based on massive conservation plus the development of 35 quads of soft energy on top of 65 quads of hard energy, this soft energy path, he says, "appears to have lower initial cost. . . ."[14] He adds, however, that "estimates of the total capital cost

143

of 'soft' systems are necessarily less well developed than those for the 'hard' systems."[15] But no matter, he in effect goes on to say in an "I just get the ideas, let somebody else work out the details" vein, the "proportions assigned to the two paths are only indicative and illustrative. More exact computations, now being done by several groups . . . involve a level of technical detail which, though an essential next step, may deflect attention from fundamental concepts. This article will accordingly seek technical realism without rigorous precision or completeness."[16] How one achieves "technical realism" without "rigorous precision or completeness" is something of a mystery. It is like designing a bridge without taking into account the stresses and strains the structure will be expected to withstand.

Noting that "Lovins has given no estimates of the total cost of his 'soft' energy path" but only a "shopping list of possibilities," Ian Forbes, president of Energy Research Group, Inc., has attempted to provide as much precision and completeness as possible in determining the total capital cost of Lovins' soft energy path for the year 2000. Forbes estimates that the total capital cost of 95 quads of energy of which 35 quads is soft energy would be $3.1 trillion in 1976 dollars, adding that the "uncertainty in this $3.1 trillion 'soft path' cost to the year 2000 is probably on the order of plus or minus $1 trillion."[17] Therefore, the total capital cost required to support Lovins' soft energy vision of the future is on the order of $2.1–$4.1 trillion, according to Forbes. By comparison, Forbes estimates that a hard energy path providing about 160 quads by the year 2000 would involve total capital costs of about $2 trillion in 1976 dollars.[18]

Total hard path capital costs of $2 trillion are thus less than even the low soft path estimate of $2.1 trillion and less than half the high soft path estimate of $4.1 trillion. In addition, the hard path produces 160 quads of energy per year while the soft path provides only 95 quads. So, on a cost-per-quad basis, the hard path turns out to have total capital costs of $12.5 billion per quad versus $22.1 billion per quad for the low estimate of the soft path and $43.2 billion per quad for the soft path high estimate. In other words, the soft path through the year 2000 would cost from 1.8 to 3.4 times more than the hard path in terms of total capital cost per quad of energy produced.

But that's not all. The soft path includes only 35 quads of actual soft energy, with the remaining 60 quads being hard energy. But soft energy accounts for the bulk of capital costs, $2.6 trillion of the total of $3.1 trillion. Applying the same plus or minus $1 trillion "uncertainty" factor to this $2.6 trillion figure results in capital costs per quad of $45.7-$102.9 billion. These actual soft energy capital costs are respectively 3.6 and 8.2 times the capital costs per quad of hard energy.

What are the chances that sufficient capital will be available to fuel a hard energy path of 160 quads or a soft energy path of 95 quads by the year 2000? Not very good, says Forbes. The estimated capital cost for the soft energy path of $3.1 trillion, for example, amounts to about 5.5 percent of cumulative gross national product at a real GNP growth rate of 3.1 percent through 2000. This is almost triple the average of 2 percent of GNP that has been spent on energy capital in the past. So also the expenditure of $2 trillion on capital for the hard path is considerably more than the historical average. Forbes points out that an expenditure on hard path capital of $1-$1.25 trillion, equal to about 2 percent of GNP, would provide about 120 quads of energy a year by 2000.[19] This would result in a moderate energy growth rate of about 2 percent a year. How much energy a year would result if the same $1–$1.25 trillion, or 2 percent of GNP, were spent on capital for the soft path? Forbes doesn't provide an estimate, but it would probably be in the range of about 70 quads a year, or only 10 more than we are producing today. This would result in an energy growth rate virtually indistinguishable from zero—a major objective of the Conserver Cult.

But how, then, is Lovins able to maintain that the soft path appears to have lower initial cost than the hard path? Forbes points out that, insofar as Lovins provides any estimates at all, he overestimates the capital costs of the hard path in areas such as coal-fired and nuclear electric power generation by 42 to 78 percent. At the same time, he greatly underestimates the capital cost of soft technology, sometimes by up to ten times or more in the case of 100 percent solar heating systems without back up.[20] For example, Lovins estimates that a solar heating system supplying all heating needs with no back up can be retrofitted at a cost of $2,700 to $3,800 for a typical home. But Forbes points out that average installed costs for such a 100 percent solar heating system based on data developed by the Department of Housing and Urban Development's Solar Demonstration Program and the Solar Utilities Company of California range from $18,600 to $34,700, not $2,700 to $3,800. These solar heating capital costs can be expected to decrease somewhat in the future, Forbes adds, but projection of major cost reductions would be speculative. Meanwhile, the figures assume that existing solar heating problems of poor reliability, low efficiency, and short lifetime will be eliminated. In addition, the figures do not reflect maintenance costs, which at this time are largely conjectural.[21]

But, even though initial costs are high, won't solar

heating eventually pay for itself since the "fuel" it uses, solar radiation, is free once the capital investment in equipment is made? This would particularly be expected to be the case if hard fuels such as oil, natural gas, coal, and uranium continue to rise in cost while cost reductions are achieved in solar heating systems. However, things don't quite work out that way.

Daniel W. Kane, a professional engineer and president of the Council on Energy Independence, has examined the total costs involved in implementing a Lovins proposal to install solar panel heating systems in virtually all U.S. homes and apartments.[22] Kane points out that a practical solar heating system normally would be designed to supply only about one-half of necessary household heating during the course of a year in most parts of the United States. This means that an additional capital investment must be made in a conventional heating system using hard fuels if "one wishes to avoid freezing at certain times during the winter." However, Kane considers only the capital costs of the 50 percent solar heating system in his analysis.

A solar heating system providing 50 percent of home heating needs costs only about one-quarter of a system meeting 100 percent of a home's heating requirements. Kane estimates a minimum cost of $7,000 for a 50 percent solar heating system. Installing this system in each home and every two apartments in the 40 million homes and 24 million or more apartments in the U.S. would result in a total cost of about $364 billion which would have to be borne by individual home and apartment owners. "There is certainly not likely to be any immediate stampede by the public to buy $7,000 solar heating systems without a heavy government subsidy, due to the simple economics of the situation," Kane observes. "Most people do not have $7,000 to spare and would have to borrow money from a bank under a home improvement loan. Such loans typically entail around 12% interest these days and require repayment in ten years. Assume that one gets a 'real deal' and the bank provides an 8-3/4% loan and a 30-year repayment schedule. The home owner then has to pay $660 per year to the bank for his solar heating system [for repayment of the loan] non-inclusive of any maintenance costs. Since the home owner will save only roughly half his gas bill, this means that his gas bill will have to exceed $1,300 per year in order to break even. Incidentally, the 8-3/4% loan over 30 years really is a 'good deal.' If the home owner gets a more common 12% home improvement loan, repayable in ten years, his yearly payment would be $1,025 and his 'breakeven' gas bill would have to be over $2,400 a year. Even with the 8-3/4% loan, natural gas prices would need to increase by three times or more before solar panel heating truly becomes barely economical to the homeowner." But, as Kane adds, if the U.S. government and/or industry were to invest $364 billion in large-scale coal gasification plants to produce synthetic natural gas for homeowner use, the "problem" of natural gas shortages would essentially be solved. And this, in turn, would effectively eliminate the need for a costly capital investment in solar heating systems in the first place.

But what about home windmills, another Lovins idea for providing energy in the future? Here, Kane points out, the costs are even more immense. He notes that a typical system is supposedly available for $10,000 to $15,000 but can supply only about 375 kilowatt-hours a month, little more than half of a reasonable estimate of electricity used in a typical home. Assuming for the sake of discussion that an adequate windmill system could be installed for

$10,000, the cost to the homeowner based on an interest rate of 8.75 percent on a $10,000 loan over thirty years would be $994 a year, exclusive of maintenance. "Unless one contemplates an electric bill exceeding $944 a year, most people are not going to rush out and buy home windmill electrical systems," Kane points out. "The cost per kilowatt-hour (kWh) from the home electrical windmill system, assuming that it generates 700 kWh/month, would be 11.3 cents/kWh. A typical current price that one pays on an electric bill is 3.5 cents/kWh. Thus, electricity would have to increase in cost-to-the-consumer by a factor of three before it is economically sound for the consumer to invest in home windmills."[23]

The economic picture for implementation of the soft energy path by central utilities instead of homeowners is hardly more encouraging. A study by the Energy Research Group of the potential for alternate power generation technologies in New England by the late 1980s came up with the following relative power costs in 1978 dollars (using base construction costs only):[24]

Technology	Technology Status	Power in Cents per Kilowatt-Hour
Solar-Thermal	Pilot	10.8 to 14.7
Photovoltaic	Research and Development	18.5 to 30.5
Wind	Developing	4.4 to 8.7
Hydro	Developed	3.2 to 14.6
Wood	Available	4.9 to 6.9
Nuclear	Available	2.1 to 2.8

The most favorable of the soft energy sources listed above—hydro—turns out to be 1.5 to 5 times more costly than nuclear power. Wood is more than 2 times costlier than nuclear power. Other soft energy sources such as solar thermal, photovoltaic, and wind are 2 to as much as 10 times more costly. And, in the case of these latter soft energy sources, the estimates not only assume a reduction in current costs but do not include the cost of back-up capacity for use when the sun goes down or the wind stops blowing.

The only soft energy source found by this study to be competitive in price for electric power production was biomass in the form of solid waste. However, this is one of the most limited of soft energy sources. It is also unique as a fuel source because it has a negative cost: cities and industries with solid waste to dispose of are willing to pay what is called a tipping fee to get rid of it. A facility using solid waste as fuel to produce electricity is thus paid to take it rather than having to pay for it.[25]

Considerable publicity has also been generated in farm areas concerning the use of another biomass fuel called "gasohol." Ninety percent gasoline and 10 percent grain alcohol, or ethanol, derived from the fermentation of cereal grains such as corn, gasohol is more costly than gasoline. It is only through government subsidies (tax breaks) that gasohol can be sold at competitive prices. In addition, production of gasohol uses more energy than is produced by the fuel. As Peter J. Reilly of the Department of Chemical Engineering and Nuclear Engineering at Iowa State University puts it, the "potential use as automotive fuel of grain alcohol (ethanol) has been a recurring issue in the American Midwest for over forty years, as it holds promise of increasing the price of these grains. Recently a second argument, that use of ethanol will reduce our requirements for energy from fossil fuels, has also been advanced. . . . [But] the production of ethanol from corn and its use in automotive fuels in blends with gasoline appears to be a poor method to raise corn prices for two reasons: the process is highly energy-inefficient, and the cost of the product is far greater than the material it is meant to replace. Though it is possible to envision situations where the waste of fossil fuel energy in producing ethanol could be greatly decreased or eliminated, even in these cases the economics are so poor that the process has little or no chance of becoming viable without tax subsidy."[26]

The one area where soft energy has made small but definite inroads is for hot water heating. But, even here, capital costs are substantial. For example, the Solar Lobby, a soft energy advocacy group, states that a "solar hot water unit costs between $420 (for do-it-yourself systems) to $2,400 (for top-of-the line equipment professionally installed)." The group goes on to say that the unit would save approximately $200 a year in oil bills, resulting in a gross return of more than 20 percent over a twenty-year period to someone in the 33 percent tax bracket.[27] But how many people, even in the 33 percent tax bracket, are going to invest several thousand dollars in a solar hot water heater in hopes of achieving savings which are by no means assured when a conventional hot water heater can be purchased for two or three hundred dollars?

The fact is that soft energy is many times more expensive on the average than hard energy, except for certain applications such as hot water heating in certain parts of the country (the South and Southwest). This may change in the future if soft energy costs come down and hard energy costs go up. But it also may not. Solar energy, for example, has been touted for the past century as a replacement for conventional fuels. When wood went up in price in the late nineteenth century, solar technology was advocated, but it could not compete with coal. The same thing happened in the 1920s when oil and natural gas proved more economical than solar energy in replacing coal. Many homes in Florida and southern California had solar hot water heaters in the thirties, but the market evaporated with the advent of cheap natural gas. And the same thing can happen again in the future, keeping solar and other soft energy sources relatively uncompetitive in cost in comparison to energy provided by hard sources such as fossil fuels, nuclear fission, and, in the future, breeder reactors and fusion.

This is why Congressman Mike McCormack (D-Wash.), who advocated aggressive, new solar initiatives as early as 1971, states that an "optimistic estimate of the total contribution that solar energy can make is 3% to 5% by the year 2000. I know this disappoints some solar enthusiasts, and I wish it were possible to produce more solar energy—sooner and cheaper. However, we would betray those who look to us for responsible leadership if we were to become 'latter-day-sun-worshippers,' ignoring the limits imposed by engineering, materials, logistics and economics, and by the sun itself."[28]

This is why Michael Noland, deputy director of the Solar Energy Research Institute, says that "it's probably heresy to say it, but we can and should be very satisfied as a nation if solar is providing seven to ten percent of our total energy needs by the beginning of the 21st century."[29]

This is why longtime solar researchers Aden and Marjorie Meinel say,

> All of us know the advantages of going solar; it is, however, necessary to review the problems that solar must surmount. We do not give you academic objections. We give you practical objections, as people who have tried most forms of solar use at

our own home (except solar cells, which we expect to try soon). We have lived with the problems, which basically are only two: 1) undependability of sunshine, especially in winter when it is needed most, and 2) high cost of the original complete installation. These two basic problems are closely followed by three corollary problems: 1) unreliability of lifetime of system components, especially the collectors themselves; 2) need for a backup energy supply, generally electrical; and 3) over-optimism in estimating performance. In spite of claims by optimists, by politicians seeking a popular cause, and by charismatic philosophers, these five problems remain in various degrees to cloud the future of solar energy.[30]

The Meinels go on to ask:

What are the real prospects for the 'soft' solar technologies advocated by Mr. Lovins? Not nearly as bright as he portrays them to be. To be sure, there are simple applications of solar energy that have long been used, such as the drying of crops and production of salt. The dream of many is that the sun will become a major energy resource, eventually to supplant fossil fuels entirely. This dream has been frustrated in the past and could be once more, as the perpetual barrier of economics is faced. . . . Mr. Lovins seems to believe that solar heating and cooling for homes and buildings is a proven and economically attractive alternative that can be used now. This just isn't the case. The fact is that solar heat is more expensive than heating by electrical resistance and far more expensive than natural gas or oil. It is hoped that costs will come down in the future, but even the most optimistic advocates talk about competitive prices based upon "lifetime costs." Most people are more interested in "first costs." If you doubt this reality, talk to any builder or manufacturer about whether the customer considers much more than first costs. . . . Mr. Lovins is certainly correct when he says that solar energy *can* be converted into heat. There are no serious technological barriers to be surmounted. No scientific breakthroughs are needed, except in the energy storage area. What defeats the effort is that cheap materials are not durable and durable materials are not cheap. . . . The problems that impede

153

the achievement of economic solar energy are the cost of collectors durable enough to survive the environment and the cost of energy storage. The operational problem is the need for solar and wind energy to have a backup energy supply, which adds to the overall cost through the existence of other expensive facilities not fully utilized during the year.

The alternative which we prefer, and which Mr. Lovins would dismiss as a "hard" technology, is the production of electric power or hydrogen on large-scale solar farms located in the arid southwest on land not now in use. Our advocacy of large-scale solar power farms is contrary to the stream of popular enthusiasm. Small-scale individual applications are the center of attention today. We feel this is contrary to the way society has gone for centuries, in fact, ever since the isolated castle of the Middle Ages gave way slowly before renewed commerce and order. There is no reason why each of us could not have our own gasoline, or diesel-powered generator and water well today—as Mr. Lovins advocates—except that it would be inconvenient, unreliable and costly. We have lived in a solar-heated house with a solar-heated swimming pool. We do not think many other persons, other than avid do-it-yourselfers, would enjoy it after the novelty wore off.

In conclusion, we would like to further disturb idealists such as Mr. Lovins by cautioning against emotional rejection of fission power as a true option for the future. The closest thing to economic reality with assured energy supplies for millenia could well be the fast breeder reactor. To foreclose this option, assuming the other options are just around the corner, could mean placing the destiny of civilization into the simple picture given by the old proverb: A bird in the hand is worth two in the bush. We are not even sure there are birds in the exotic energy bush. We perceive the bush to be moving, but it could only be the gentle wind mocking us.[31]

There are some other solar energy advocates who are not averse to admitting that this form of energy is costly. Solar energy pioneer Howard Odum, for example, points out that "materials, fuels, and labor are fairly high quality energy, whereas solar energy is low quality. Energy use is maximized when relatively little high quality energy is required to process and obtain low quality energy. In solar technology the ratio of high quality energy to low quality sunlight is higher than in some other alternative investments."[32] Solar energy writer William Shurcliff states that "new inventions are needed because most of the solar heating systems now being produced are too expensive and too complicated. . . . Today more than 100 companies are making and selling collector panels. . . . Yet the complexity and cost are often frighteningly high."[33]

This, of course, is in contrast to Conserver Cult attempts to portray solar and other soft energy technologies as being here right now in terms of economics. But, even though the cult promotes this view, its real position more often appears to be that costs are simply irrelevant from the standpoint of achieving the greater good embodied in its vision of the low-energy, low-technology, decentralized soft society of the future. For example, in his introduction to *Small Is Beautiful* Theodore Roszak states that the deliberate intention of author E. F. Schumacher is to "subvert 'economic science' by calling its very assumptions into question, right down to its psychological and metaphysical foundations."[34] Schumacher, he says, provides "a nobler economics that is not afraid to discuss spirit and conscience, moral purpose and the meaning of life, an economics that aims to educate and elevate people, not merely to measure their low-grade behavior."[35] Presumably, "low-grade behavior" consists of considering the costs involved in achieving psychological and metaphysical elevation.

Lovins observes that "however well economic comparisons are done, serious questions can be raised about their relevance today. Even Keynes admonished us not to 'over-state the importance of the economic problem, or sacrifice to its supposed necessities other matters of greater and more permanent importance.'" Lovins approvingly cites the statement of Alan Poole that "the ultimate condemnation of a project is that it is 'uneconomic,' which is to say it costs 5 mills per kW-h(e) more than another option. By comparison, human labor in a dirt-poor pre-industrial society costs roughly 2000 mills/ 0.5 kW-h(e), 4,000 mills per kW-h(e). Does it *really* matter if, say, a solar option costs twice as much as a nuclear option?" Lovins sums up by noting that "since human labor is three orders of magnitude more expensive than inanimate energy, very substantial increases in the comparatively negligible price of the latter may be of no great moment."[36] In other words, as long as you've got more energy than people in a "dirt-poor pre-industrial society," it's unchic to worry about what energy costs.

But solar advocate Stephen Lyons manages to do the best job of turning the subject of energy costs as well as modern economic history upside down, stating that "preoccupation with costs, especially when it's pretended that they can be given absolute numerical values, obscures the important principle that societies choose technologies for noneconomic reasons, then adopt whatever economic policies are needed to make them float . . . ; the most important step toward a solar future is to debunk the widespread belief that our social decisions are determined by economic factors."[37]

This is the ultimate statement of the Conserver Cult's approach to costs: its vision of the social and political future of America is right and therefore should be achieved *regardless* of cost. However, much as the cult rails about "debunking" the ever-present and inescapable reality of energy costs, it recognizes that costs are a barrier that must somehow be surmounted if the soft path is to be followed. One way is to "reduce" the cost of solar and other soft technologies through government subsidies. This is the approach taken by the Solar Lobby, which promotes a massive government effort involving an expenditure of $50 billion on soft energy research and development and "an ambitious and wide-ranging set of incentives, tax credits, loans, guarantees, demonstration programs, job training programs and other positive inducements to the private sector to make solar choices."[38] Apparently, President Carter has bought much of this approach because he has proposed a package of solar energy programs and set a goal of meeting 20 percent of U.S. energy needs by the year 2000 from solar and renewable resources. Speaking on the White House roof in front of the presidential mansion's newly installed solar water-heating system, Carter declared that "no one can ever embargo the sun or interrupt its delivery to us."[39] He did not mention Whoever it is that causes night to fall. He also did not mention that fossil fuel and nuclear research and development programs were being cut back to make way for the solar energy package.

White House officials estimated that the entire package, if passed by Congress, would cost up to $3 billion through 1985.[40] However, what if, even with all of these subsidies, incentives and inducements, people still find solar energy uneconomical and therefore don't make "solar choices"? Most people would begin to conclude in this case that perhaps solar energy is not all that it's cracked up to be. But not the Solar Lobby, which is prepared for this eventuality, recommending that "if by 1985 a series of benchmark goals have not been met, a standby *mandatory* program would go into effect [emphasis added]." [41]

Another even more effective way of force-feeding the Conserver Cult's low-energy, low-technology future to the American public is to use government power to restrict production and increase costs of conventional domestic energy so that the high-cost soft path becomes increasingly more "competitive." This is the approach the cult has taken through ever-more stringent environmental laws, absolute-safety regulations, and comprehensive and complicated price controls.

11

ENERGY
AND THE
ENVIRONMENT

Energy is the capacity to do physical work. The use of increasing amounts of energy to do physical work has spurred the socioeconomic development of people throughout history. It has also provided a human environment which is more hospitable to people's health than the natural environment. Living in the natural environment thousands of years ago, people had a lifetime, or longevity, of no more than twenty to thirty years while the vast majority of babies born did not survive to adulthood. Today, in advanced countries, average life spans are seventy years or more, and close to 100 percent of infants live to become adults.

Studying the increase in longevity and reduction in infant mortality which has occurred in more recent times, the International Institute for Applied Systems Analysis in Laxenburg, Austria, found that the two variables most closely associated with these improving health conditions were per capita energy consumption and literacy. Both per capita energy consumption and literacy, the institute concluded, show "strong interactions with longevity" and a "strong and predictive relationship to declining infant death rates."[1] The institute provided no explanation of this relationship, but one obvious one would be that increased use of energy freed people in general and mothers in particular from the back-breaking and debilitating physical work of the past, aside from other energy benefits such as increased warmth and protection.

The increasing use of energy thus provides people with an improving human environment in terms of both continuing material advancement and better health conditions. However, these benefits in the human environment are achieved at a cost to the natural environment. This is because producing energy, like doing physical work, is a dirty business. People sweat and get dirty when they do physical work. So also people "dirty-up" the natural environment when they produce energy to improve the human environment.

Throughout history, the benefits of energy production to the human environment have more than outweighed the costs of degradation in the natural environment. Otherwise, socioeconomic development would not have occurred side-by-side with improving health conditions. But this has only been because a reasonable trade-off has been continually achieved between energy production and adverse environmental effects. Take, for example, the caveman who first discovered how to harness fire. Living in the natural environment, he resided in a cave which was not only dank and dark but dangerous, since predatory animals could enter at any time. The discovery of fire presented him with the opportunity to improve his human environment by using fire to provide heat, light and protection. However, the fire also produced smoke, which polluted his environment. Locating fire inside his cave probably would have resulted in quick asphyxiation. Locating it outside the cave would have nullified the fire's benefits. So the caveman achieved an acceptable trade-off between his human and natural environments by locating the fire at the mouth of the cave. Undoubtedly, smoke from the fire still befouled the cave when the wind blew the wrong way. But it seems safe to assume that the caveman looked on an occasional lungful of smoke as a small price to pay for the benefits of heat, light and protection provided by the fire.

It is just such positive trade-offs as this between the benefits of an improved environment created by the increasing production of energy and the costs of the polluted natural environment resulting from increased energy consumption that have spurred human advancement in the past. Of course, such trade-offs always run the risk of turning negative over time with continuing growth, that the costs of pollution can become greater than the benefits of energy use. However, the solution to this problem is not to turn the clock back to some supposedly idyllic previous era of lower energy consumption. *There is no such era.* The historic answer has always been to recreate a positive trade-off between the human and natural environments through *increased* energy production and the development of new, less-polluting energy sources.

Pollution and degradation of the natural environment are hallmarks of all civilizations; they in effect come with the human franchise. But, contrary to Conserver Cult fantasies, they are even more severe problems in low-energy, low-technology cultures than in high-energy, high-technology societies. Take, for example, the American Indians, who had the epitome of a low-energy, low-technology civilization. The Indians are oftentimes idolized by the Conserver Cult as people who lived in harmonious balance with their natural environment. It is, of course, true that many Indian tribes practiced a form of pantheism in which the natural environment was held to be sacred. But this reverence for nature did not prevent them from also practicing an environmentally destructive form of agriculture which is characteristic of most low-energy, low-technology societies. This agricultural method basically consisted of burning off a forest or prairie in order to instantly fertilize the land with ashes from the fire, making it amenable to planting. However, nutrients added to the soil by the fire were exhausted after a few years of planting, requiring that additional forests and prairies be continually put to the torch.

Like other primitive peoples, the Indians used this agricultural approach precisely because it conserved on the need for energy—in this case, their own human energy—and technology. But its effects on the natural environment were devastating. Dr. James R. Dunn, an eminent geologist and chairman of the board of Dunn Geoscience Corporation, comments that it is "not generally realized that the Indians burned woods and prairies extensively. Burnt Hills, New York, got its name because the Indians regularly burned the woods. . . . The Indians created much of our prairie, probably well over 100,000,000 acres. . . . By burning prairie and exposing bare soil, they created dust storms such as that described in Kansas in 1832, *well* before the plow and settlers."[2]

How much land would we need if we were "environmentalists" like the Indians? Dunn points out that "if 400,000 Indians destroyed over 100,000,000 acres of woodlands, 215,000,000 would have destroyed 52,500,000,000 acres, about 25 times the whole present United States, i.e., the forests of all North America would have vanished long ago."[3]

But the fact is that our population did grow to more than 200 million and yet the forests have not only failed to vanish but are larger today than they were when the first European set foot in North America. How did this happen? It happened as a result of the increased use of energy in the form of animal power to pull a new type of technology—the plow. The animal-drawn plow made it possible to use the same fields over and over again, eliminating the need to burn forests and grasslands to obtain new, fertile fields suitable for planting. The increased use of energy eliminated the need for environmental degradation. As the world-renowned conservationist John Muir put it, where "an Indian required thousands of acres for his family, these acres in the hands of industrious, Godfearing farmers would support ten or a hundred times more people in a far worthier manner. . . ."[4]

Or turn the clock back to another relatively low-energy, low-technology society at the start of the Industrial Revolution: England, specifically the city of London in the year 1661. Diarist and Royal Society Fellow John Evelyn described the air in London at the time as an "impure and thick Mist, accompanied with a fuliginous and filthy vapour, which renders them obnoxious to a thousand inconveniences, corrupting the Lungs . . . causing Catarrhs, Phtisicks, Coughs and Consumptions [to] rage more in this one City than in the whole Earth besides." The cause, explains Prof. Petr Beckmann, was the burning of coal, which the English then called "sea coal." Pollution by coal smoke had actually been troublesome in England since the end of the thirteenth century. But this did not stop growth in its use because improvements in the human environment resulting from coal burning outweighed the costs of pollution in the natural environment. During a coal shortage, for example, Robert Gesling reported in a pamphlet published in 1644 that "some fine Nosed City Dames" used to complain about "the Smell of this Cities Seacoale Smoke," but now they cry "would to God we had Seacoale. Oh the want of Fire undoes us!" As Beckmann relates, London went on to become the world's leading industrial city. But it also became the city of the famous London fogs which, until the 1950s, had people groping along walls in zero visibility at high noon. However, in 1952, the coal-burning costs of polluting the natural environment more than caught up to the benefits in the human environment when 3,900 Londoners died within a week after a particularly bad pollution episode. But the solution to the problem was not to ban coal and retrogress to the twelfth century. It was to reestablish the positive trade-off between human and natural environments by using more energy to operate pollution control equipment. This banished London fogs for good, while further prolonging the life of the average Londoner by several years.[5]

But it is not really necessary to turn the clock back to observe that the positive contributions to the human environment made by the increasing use of energy far outweigh the negative costs of pollution in the natural environment or that the problems experienced in the natural environment by high-energy, high-technology societies are far less serious than those in low-energy, low-technology cultures. As Dunn puts it, we should be continuously reminded that

> only the wealthy can afford sewage systems, sewage disposal plants, and garbage disposal. . . . Only wealthy techological societies can afford to leave steep slopes in woods and till or pasture only flat or gently rolling lands; only the wealthy can control erosion to the maximum; only the wealthy can afford forests. A look at the less developed countries of the world discloses a dismal picture of environmental degradation on a massive scale. . . . Not that we technologic nations do not have pollution problems that require action. But, remember, we often measure environmental problems in parts per million and worry about what could conceivably happen in decades or centuries; conversely, undeveloped nations see environmental problems in terms of what is happening; in terms of starving children, short life spans, stripped forests; the high angle of tilled slopes, the number of sand dunes, expanding deserts, loss of water resources, etc. These are things that we are rarely aware of, things that we have been sheltered from for so long that we have forgotten that they existed.[6]

There always is a trade-off between the benefits to the human environment resulting from energy production and the costs to the natural environment caused by energy consumption. With growth in energy production and consumption, this trade-off becomes increasingly more positive. However, the Conserver Cult ignores improvements in the human environment resulting from growing energy production while singlemindedly deploring pollution in the natural environment resulting from increased energy use. The only trade-off acceptable to the cult is one in which the natural environment is protectd from being sullied by human hands at any cost, even that of a catastrophic energy shortage. This approach to environmental issues began in the sixties, a time of unparalleled affluence when, as John Whitaker notes, the "combination of spare money and spare time created an ambiance for the growth of the causes that absorb both money and time. Another product of affluence has been the emergence of an 'activist' upper middle class—college-educated, well-heeled, concerned, and youthful for its financial circumstances. The nation has never had anything like it before. It is, in fact, a conflict in terms—a mass elite. It is sophisticated, resourceful, politically potent, and compulsively dedicated to change, to 'involvement.' It forms the backbone of the environmentalist movement in the United States."[7]

The "mass elite" described by Whitaker also has one other distinguishing characteristic. It is far removed from the work-a-day world in which energy production and environmental pollution are inextricably and necessarily intertwined. All of us, rich and poor, share a common natural environment—a commons, so to speak—consisting of the air, the water and the land. Improvements in the human environment inevitably cause pollution in this natural commons. This is understood and accepted by people at the lower end of the economic ladder because the environmental degradation is more than compensated for by improvements in their human environment in the form of better jobs, higher income, larger houses and greater mobility. The worker in the steel mill understands that his job depends to some extent on producing air pollution. The employee in the chemical plant accepts the fact that his income is partly derived as a result of water pollution. The construction worker is well aware that his livelihood involves disturbing the land. But the mass elite which is at the higher end of the economic ladder has none of these perceptions. Ensconced in a comfortable human environment which already meets most of their needs and

166

engaged primarily in intellectual and organizational endeavors unrelated to physical work, they look on energy production as a crass reminder of a previous benighted period which is now beneath them. Environmental pollution therefore is viewed as a completely unwarranted invasion of *their* commons, which must be repulsed at all costs, including the sacrifice of the opportunities, livelihoods and futures of other people who are below them on the economic and social ladders— in other words, most of the human race.

The environmental movement began as a grass roots cause but did not long stay that way. "It was a citizen movement, not a professional or governmental one," states David E. Lilienthal, founder and chairman of Development and Resources Corporation, and former chairman of the Atomic Energy Commission and the Tennessee Valley Authority. But Lilienthal adds that, "unfortunately, I think environmental protection is in danger of becoming professionalized and institutionalized—indeed, to a considerable extent anyone who observes the Washington scene these days can see this has already happened. From being a citizen, volunteer effort scattered in thousands of communities, paid professionals have more and more moved in and are set to capture the environmental movement, take over its direction, install sophisticated methods for the raising of funds, dictate the expenditure of those funds, and become the self-appointed spokesmen and politicians of conservation."[8]

What this professionalization and institutionalization means in practice is revealed by the Heritage Foundation, which reports estimates that the environmental movement today encompasses some 3,000 organizations, of which several hundred are national or regional in scope.

Twelve of the largest of these groups enjoy the support of more than 4.3 million people and have combined annual budgets of more than $48 million. Support for the larger organizations is derived in significant measure from grants made by government and tax-exempt foundations such as the Rockefeller Brothers Fund, the Andrew W. Mellon Foundation, and the ubiquitous Ford Foundation.[9] Among the more prominent environmentalist organizations are Friends of the Earth, National Resources Defense Council, Environmental Defense Fund, Sierra Club, Wilderness Society, Conservation Foundation, National Wildlife Federation, Environmental Action and Worldwatch Institute. Among these and other environmental organizations there exists a considerable overlapping of interests, as well as extensive cooperation on matters of mutual concern. There is also a pattern of interlocking relationships through shared supporters, officers and board members who are primarily drawn from government, universities, foundations, the literary and entertainment world and the social register.

With the advent of the Carter administration, the movement also began to serve as a conduit for placing environmental activists in key administration posts. One of the big winners in this regard was the National Resources Defense Council. Gus (J. Gustave) Speth, who had been with NRDC since its founding in 1970, became one of three members of the President's Council on Environmental Quality, which is the president's chief advisory body on environmental matters. John Leshy, an NRDC attorney who specialized in forestry, coal-leasing and public-lands issues, was appointed associate solicitor for energy and resources in the Interior Department. NRDC's David Hawkins became assistant administrator for air and waste management in the U.S. Environmental Protection Agency, or, as NRDC proudly put it, the "chief air pollution officer of the United States." Richard Hall, who had left NRDC but continued to work closely with the environmental organization, was appointed a special consultant to the Interior Department's Surface Mining Project.[10]

What is the energy philosophy of the environmental organization from which these staff members were propelled to positions of power within the Carter administration?—the basic Conserver Cult approach that "there is a better path to follow" which will lead to what NRDC executive director John H. Adams calls "a future of slow energy growth, tight conservation measures, and research and development of energy 'income' technologies such as solar, geothermal and wind."[11]

Based on the conclusion of the International Institute for Applied Systems Analysis that improving health conditions are most closely associated with per capita energy consumption, it would seem that slowing down energy growth would be a way of creating worse environmental conditions than would otherwise be the case. It would also seem appropriate to take into account that a shift has already naturally begun taking place among hard energy resources from coal, which, according to a health evaluation conducted by the American Medical Association, is the most environmentally damaging, to oil, which is less damaging, to natural gas and nuclear power, which are the least environmentally damaging energy sources of all.[12] However, such common-sense distinctions find no place in the Conserver Cult meat-axe approach to energy production of summarily turning off growth in all hard energy production while force-feeding the development of the supposedly environmentally benign energy provided by soft sources.

The first and most important method used by the environmentalists to thwart hard energy production is to deny energy producers the right to lease areas containing fossil fuels. This is possible because the federal government owns more than 50 percent of land and water areas containing hard energy resources, including public land in the West with about 85 percent of strippable, low-sulfur coal, all of the Outer Continental Shelf (OCS), where offshore oil and natural gas are located, and onshore areas with 80 percent of high-grade oil shale.[13] How this federal control of the people's land has been used by environmentalists to tie up the leasing of coal is most instructive.

For all practical purposes, the government coal-leasing program has not been in operation since 1971 due primarily to environmentalist pressures. Meanwhile, the vast majority of existing, unworked coal leases have become unmineable as a result of environmental legislation. H. Peter Metzger points out, for example, that in seven western states there are currently 519 existing and unworked federal coal leases. But he adds that an examination by Charles Margolf of W. R. Grace and Company showed that 95 percent, or 491, of these leases were unmineable—because they were all issued before 1970 and the passage of a veritable flood of environmental legislation, including the National Environmental Policy Act, the Clean Air Act of 1970, the Clean Air Act Amendments of 1977, the Clean Water Act, the Clean Water Act Amendments, the Coal Leasing Act Amendments of 1975, the Surface Mining Control and Reclamation Act of 1977, the Critical and Endangered Species Act, the Federal Water Pollution Control Act Amendments of 1972, the Mine Safety and Health Act, and the Mine Safety and Health Act Amendments of 1977, not to mention other federal and state legislation and a host of rules and regulations of numerous agencies of the federal government. Obviously, as Metzger observes, the laws, rules and regulations applicable to coal mining today are considerably different from those which existed prior to 1970 when 95 percent of the existing leases were issued.[14]

But why haven't new coal leases been issued since 1971? It was in 1971 that the U.S. Department of the Interior declared a temporary moratorium in coal leasing to study whether its leasing program was spurring coal production. The moratorium ended in 1974, but Interior made only ten lease sales between 1974 and 1977, while performing various reviews, some of which were challenged in court by the environmentalist NRDC.[15] Then, on September 27, 1977, U.S. District Court Judge John Pratt ruled against Interior and in favor of NRDC, which had brought suit in *NRDC* v. *Hughes* to force Interior to rewrite and greatly expand an already-prepared environmental impact statement on coal leasing. The NRDC suit was concerned with the supposedly irreversible damage which strip mining would cause in arid western lands, even though the Surface Mining Control and Reclamation Act passed in 1977 requires that producers restore strip-mined land to its original shape and use. The court decision effectively froze issuance of new coal leases until at least 1981.[16]

But even more shocking than the resulting extension to ten years of the hiatus in new coal leasing was the fact that the court case of *NRDC* v. *Hughes* was in Metzger's words *"definitely* not an adversary proceeding" in the traditional legal sense but rather a "formalizing of plans for long-range national coal development made between individuals of *identical and rather extremist philosophy. . . ."* Metzger explains that

John Leshy is at least one Department of Interior lawyer who is preparing defendant Interior's Environmental Impact Statement as demanded by plaintiff NRDC. [But Leshy] was actually employed until last year as the NRDC lawyer who prepared plaintiff NRDC's case against defendant Interior. In other words, Leshy is "defending" against *the very case he created.* Also representing Interior in its so-called "defense" against NRDC's suit was U.S. Justice Department Assistant Attorney General James Moorman, who took that post after leaving his job as executive director of the Sierra Club legal defense fund. To add to the non-adversary nature of *NRDC* v. *Hughes.* Sierra Club attorney Bruce Terris took over as NRDC's attorney when Leshy left NRDC to work for defendant Interior. Thus, in *NRDC* v. *Hughes,* we see plaintiff's attorney Terris arguing "against" defendant's attorney Moorman when just a year before both men were Sierra Club lawyers.[17]

With such a cozy relationship, says Metzger, it's no wonder that another "remarkable development" took place in which the U.S. Department of the Interior, in effect, gave up its decision-making power on federal coal leasing to NRDC. Metzger explains how this happened:

Since [U.S. District Court Judge John] Pratt's decision effectively stops coal development in the U.S. cold, the Department of Interior appealed *NRDC* v. *Hughes.* But NRDC, fearing a reversal, was willing to compromise. Using the excuse of the sure delay which would have been caused by a lengthy appeal, the U.S. Department of Interior agreed not to appeal the case in return for NRDC agreeing that a certain limited amount of coal leasing can commence. Thus, Interior can only release those federal lands for coal leasing that NRDC approves of in advance. In other words because NRDC has a *veto* over any leases proposed by Interior, *NRDC, not Interior, now controls Federal coal leasing policy in the United States. . . .* So here's what has happened finally: Under the pretense of an adversary proceeding, NRDC and Sierra Club attorneys, some still employed by NRDC and some now in the Interior Department, have conspired together and succeeded in moving the authority to make national coal leasing policy from the government to a private environmentalist pressure group.[18]

In June 1979, Interior Secretary Cecil Andrus announced that the U.S. planned to lease rights to 1.5 billion tons of federal coal in 1981 and 1982. Stating that the federal coal leasing program "responds to President Carter's directive that we provide a rightful place for federal coal in the nation's energy future," Andrus said the program would contain safeguards to make it "environmentally acceptable" by requiring preparation of regional environmental impact statements in 1980, with leasing to begin in 1981.[19] Andrus did not point out that leasing was not scheduled to begin until after the next presidential election, nor did he add the NRDC was in a position to nix any proposed leases.

But the conclusion that the federal coal leasing program is an environmental clinker from the standpoint of increasing coal production comes through unmistakably in comments by the National Coal Association that the program "represents an unprecedented degree of management and control at the Departmental level of federal coal resources. Many specific elements of the Program are unworkable. If implemented, we believe it would be impossible for the Department or the private sector to achieve in timely or responsible fashion any realistic goal of resumed federal coal leasing. The unworkable aspects of the Program appear in virtually all respects to be actions or choices within the discretion of the Department. They are not mandated by external constraints of law or national policy. As a result, the Program would appear to represent the conscious adoption by the Department of a land management policy which is systematically biased against federal coal development."[20]

So much for coal leasing. What about leasing for oil and natural gas exploration and development? About 95 percent of the Outer Continental Shelf, which contains the greatest potential for significant new discoveries, has never been explored for oil or natural gas. The major fear of offshore drilling expressed by environmentalists is that oil spills will "irreversibly" pollute the marine environment. Yet the facts are that, contrary to emotional, doomsday predictions following the Santa Barbara oil spill in 1969, offshore oil pollution is a nuisance but there is not much hard evidence that spills have any long-term effect on marine life.[21] In addition, a 1974 study by the National Academy of Science blamed offshore drilling for no more than 0.9 percent of marine oil pollution, as opposed to 15.8 percent from tankers and 35.1 percent from river runoff.[22]

One of the major areas promising oil and natural gas finds is the Baltimore Canyon in the Atlantic Ocean. But environmental suits barred the Interior Department from accepting bids on leases in this area until August 1976.[23] Leasing then proceeded, but at a slow pace and under heavy environmentalist pressure. NRDC tells in its 1977–78 annual report how the environmentalist organization has "taken a leading role to ensure that OCS development proceeds with adequate environmental safeguards. NRDC's initial success in challenging the first Atlantic OCS lease sale was reversed on appeal. However, the months of proceedings and attendant publicity influenced the Department of the Interior to adopt new procedures which significantly increase environmental controls over OCS leasing and address many of NRDC's original objections. Most importantly, these include a full-scale environmental assessment after discovery of offshore oil to determine whether and how development should proceed, and Interior's adoption of authority to suspend drilling operations which threaten to cause significant environmental harm."[24] In addition, in the Atlantic's Georges Bank area, environmentalists have joined with the State of Massachusetts to demand performance of predevelopment environmental baseline studies that one marine biologist estimates would take seventeen to twenty years to properly complete.[25] Hardly calculated to inspire confidence in potential lease developers, this environmentalist ring-around-the-rosie of "now you have a lease, now you don't" gives an extremely hollow ring to the Interior Department's proposals to "step up" the sale of offshore oil leases from March 1980 through February 1985.[26]

Another highly effective way of squelching the production of hard energy is simply to declare land off limits to mining and drilling. One way environmentalist organizations accomplish this is by getting land set aside as "wilderness," which is defined from a pure environmentalist point of view in the Wilderness Act as a place where "earth and its community of life are untrammeled by man, where man himself is a visitor who does not remain." The federal government owns about 761 million acres of land, which is primarily in the West and Alaska and amounts to about one-third of the country. Although it contains about half of the country's known energy reserves, this land contributes only 10 percent of the nation's energy production. Up until 1968 most of this federal land was open to development. But now, as a result of environmentalist pressures, most is closed or restricted. One survey, for example, showed that only 17 percent of federal lands were off limits to oil and gas development at the start of the 1970s but that more than 60 percent is closed or restricted now. Environmentalists such as Michael Mc-Closkey, executive director of the Sierra Club, say this withdrawal of land from energy production is vital in order to identify and protect areas before they are gouged up by mines or dotted with oil wells. But William Dresher, dean of the college of mines at the University of Arizona takes a different view, stating that "it's as if some superior being was attempting to stifle the future growth of the U.S. deliberately."[27]

A significant portion of the public land being studied for conversion to wilderness or other restricted uses falls under a U.S. Forest Service study called RARE II (a second version of the first Roadless Area Review and Evaluation). It should come as no surprise that wilderness continues to expand rapidly when one learns that the RARE II director is Carter appointee George Davis of the Forest Service—formerly executive director of the Wilderness Society. In addition, final RARE II responsibility belongs to M. Rupert Culter, assistant secretary of agriculture, who previously served as a high official in the Wilderness Society.[28]

But at least the American people will be able to benefit from all of this newly created wilderness, right? Wrong! The environmentalists mean it when they say that these areas are to be "untrammeled by man." Alaskan and former Secretary of the Interior Walter J. Hickel explains that

the public is being told that great national parks are being created that they will one day be able to visit. But there are no plans to provide access to these areas so that average Americans of all classes, senior citizens, the very young, or the disabled can have a chance to view and enjoy the wonders of these areas. An exclusive domain is being created for the bureaucrats and the wealthy elite, and at tremendous cost. During the next few months, the National Park Service will hire 110 more employees, at the expense of $5 million a year, to police these areas *to keep the people out*. The public thinks this land is being set aside for them. They will learn differently when they run into 'no trespassing' signs and uniformed officials [emphasis added].[29]

The supposed threat to so-called endangered species is another gun in the hands of environmentalists shooting down proposed energy projects. The Alaska oil pipeline, for example, was vehemently opposed by environmentalists, on the grounds that it would have a destructive effect on animals such as the caribou. To placate the environmentalists, expensive "gates" were engineered and the mostly aboveground pipeline was buried at points where it crossed the migratory paths of the caribou. But it turned out that the caribou didn't appreciate what was done for them. They actually like the pipeline aboveground. They sleep under it, play under it, leap over it and enjoy its friendly warmth.[30]

However, aside from the question of whether any proposed energy project will have the destructive effects envisaged, environmentalists claim that the natural habitats of endangered species must be preserved because man cannot eliminate a single species without potentially affecting the whole tightly knit fabric of life. Ignoring the fact that millions of species have been eliminated by the natural environment itself over past millennia, Congress enacted this environmentalist view into law in the 1973 Endangered Species Act, which now lists 600 species of animals, fish, reptiles and plants—with another 2,000 under consideration for listing.

In 1978, construction of the Tellico Dam, which was 95 percent complete at a cost of $113 million, was stopped cold by the U.S. Supreme Court because it endangered

177

the natural habitat of the snail darter, a three-inch-long fish which had not even been discovered until 1973. In another case, construction of a dam in Maine was suspended because it might result in the destruction of the furbish lousewort, a small plant with fernlike leaves and yellow flowers.

It is now almost a matter of course for environmentalists to come up with what Metzger calls an "endangered red-herring" on any energy project which they oppose. As Sen. James McClure (R-Ida.) comments, unless the Endangered Species Act is amended to allow the balancing of environmental with energy and economic issues, it will be used as a "weapon" to stop important dams and other projects.[31] Congress responded to this situation by creating an Endangered Species Committee empowered to exempt government-financed projects from the law if no feasible alternative exists and the project's economic benefits outweigh those of other causes.[32]

However, this has not changed the basic environmentalist policy of placing animal, fish, reptile and plant interests above human concerns. One of the most publicized of endangered species, for example, is the whale. Three or four whales eat about as many fish in a day as are daily destroyed by the cooling water intake of a nuclear powerplant. Environmentalists opposed construction of a nuclear plant in Seabrook, New Hampshire, because of this fish kill but are simultaneously operating a worldwide campaign to save whales.[33] In other words, it's okay for whales but not okay for human beings to obtain energy by killing fish.

The same environmental extremism so effectively used to prevent energy projects from being developed in the first place is also applied in the prevention of energy production and use. Milton Copulos, policy analyst at the

178

Heritage Foundation, points out, for example, that the standards of the Clean Air Act are "so stringent that they are impossible to meet, at least in some categories."[34] They are so tight, in fact, that even the natural environment can't meet them on occasion. This was the case with a shale oil project that Standard Oil (Indiana) tried to develop in Colorado. The company had to shut down operations because, as board chairman Swearingen relates, it "wasn't that *we* could not meet [the standards]— the culprit was Mother Nature herself. Because of the emissions of plants and wind-borne particulates, the natural quality of the air could not meet federal standards."[35]

As might be suspected, Environmental Protection Agency air quality standards are based on research that is oftentimes dubious at best. For example, the EPA used the winning times of California high school cross-country runners as the prime criterion for deciding what level of hydrocarbon emissions should be permitted from automobiles, a rather "weak reed," as the *Wall Street Journal* put it, on which to "pyramid billions of spending by the auto industry.[36]

Here's another example: Based on a 1972–73 report called *CHESS,* the EPA set stringent limits on the amount of sulfur dioxide that may be emitted by coal-burning powerplants, even though there was evidence that industrial workers had been exposed to sulfur dioxide for many years at hundreds-of-times-higher concentrations with no apparent adverse health effects. It is estimated that it will cost consumers (you and me) $200 billion between now and the end of the century to install EPA-mandated flue-gas scrubbers in the stacks of coal-burning powerplants to meet the agency's standards for sulfur removal.[37] This expenditure on scrubbers will be exorbitant in the extreme from a health point of view, according to Irwin W. Tucker, professor of environmental engineering at the University of Louisville. Tucker points out that the EPA standard is ridiculously strict. Under it, the amount of sulfur that would be breathed in by an average individual over a period of several years would be no more than that ingested by drinking a single bottle of pop or lemonade.[38] The scrubber expenditure will also be more than has to be spent because the EPA standards can be met at a cost of five to eight dollars per ton of coal by partially desulfurizing coal or blending low- and high-sulfur coal instead of installing scrubbers at a cost of fifteen dollars per ton of coal.[39] Furthermore, according to EPA's own figures, the actual concentration of sulfur dioxide in the air has been decreasing to the point where today it is less than 60 micrograms per cubic centimeter, 35 percent less than in 1960.[40]

Why, then, is the EPA insisting on the use of scrubbers for all coal-burning plants, even those using low-sulfur coal that does not require scrubbing to meet EPA standards? Eugene Guccione, editor of *Coal Mining & Processing,* notes that the "answer was given . . . by Sierra Club field representative Dave Gardiner who called the EPA plan 'an economic equalizer . . . that will equalize the cost of pollution control all over the country.'"[42] But, in addition, the plan will seriously reduce financial incentives to mine low-sulfur coal in the West, bringing closer the environmental goal of zero western coal mining which is dear to the heart of the Sierra Club. It is for reasons such as this that William N. Poundstone, executive vice president of Consolidated Coal Company, observes that

a very important question we ought to be asking ourselves is:

"What is the health effect on people of limited income when foolishly imposed sulfur dioxide clean-up costs drive their monthly energy bills beyond their means?" Consider, as an example, an elderly widow living on a $200 Social Security check each month, with electricity and fuel bills averaging $50 a month. If the costs of installing flue gas scrubbers or buying fuel of needlessly low sulfur content drive her energy bill up by 20 percent—which it could well do—it becomes a problem of major proportions for her. It means $10 less per month for food, for medicine, for clothing, for everything she needs to live.[42]

The same story of environmental extremism can be told in relation to other energy projects. Petr Beckmann relates how in early 1975 Dow Chemical Company "proposed a $500 million oil refinery near San Francisco. Pollution control authorities admitted it was the cleanest facility they had ever seen, but denied a permit anyway, with the cryptic explanation that no permit could be granted even if Dow were to meet all emission standards. After two years' effort and $10 million in costs, Dow could obtain only four of the required sixty-five (!) permits. The rest were not rejected—Dow was 'simply unable to hack through the regulatory morass and get straight answers.' Early [in 1977], therefore, Dow cancelled the project."[43]

Standard Oil Company of Ohio abandoned plans to build the 1,000-mile PACTEX pipeline to transport oil from the West Coast to eastern and central states because of "killing" delays over nearly five years in obtaining 750 federal and state clean air and other permits.[44] "I'm afraid new, major energy projects in the United States have little hope of success today," commented SOHIO chairman Alton W. Whitehouse. "A quagmire of federal and state regulations now exists that can bog down any project, no matter how worthy and regardless of the national interest.

181

The lesson of PACTEX is that no major energy project can be seriously considered in many areas of the country by industry today unless governmental processes are changed to provide answers within reasonable time periods."[45]

Environmentalism in the United States has resulted in what is called the American joke, which would be funny if it weren't so close to the truth. It goes like this: Why does it take five Americans to change a lightbulb? Answer: One to turn the bulb and four to file the environmental impact statement. More to the point, Peter J. Brennan, chairman of the New York State Committee for Jobs and Energy Independence, declared that "what worries me is that the good and proper environmental protection laws and procedures are being perverted and exploited by a gang of 'no-growth muggers.' To the working people of this nation, our offshore energy resources are a matter of bread and butter. Many so-called environmentalists want us not to use these energy resources. They are saying, 'Let the public eat cake,' or even 'pie in the sky'—insisting that conservation and energy schemes like solar power, windmills and such can meet our energy needs in the years now facing us."[46]

But, even if soft energy resources such as solar and wind power can't meet our needs, they will at least be environmentally benign, won't they? Nope. Douglas Martin reports that "researchers are finding that solar technologies and other self-renewing energy sources also pose environmental problems, ranging from toxic chemical pollution to waste disposal to just plain ugliness. These novel new energy sources, like conventional fuels, carry ecological costs." Martin goes on to say that even Denis Hayes, originator of Sun Day and newly appointed director of the Department of Energy's Solar Energy

Research Institute, concedes that "certainly, it's something that people should be thinking about."[47]

What are some of the adverse ecological effects of soft energy production? First of all, soft energy gobbles up huge amounts of land. Centralized solar power facilities require more than twelve times as much land as, for example, nuclear power plants. Windmill installations directly use more than sixty-six times as much land as nuclear powerplants and 2,300 times more in terms of total land affected from the standpoint of visual impact as well as use.[48] Secondly, solar and wind centralized energy production systems use materials-intensive technologies that require environmentally hazardous extraction and production operations.[49] Finally, even the simplest solar devices used to heat and cool homes are potentially scarred by environmental hazards. According to Martin, government experts, for example, fret that the antifreeze and other toxic fluids used in solar collectors might escape into drinking water, contributing to such diseases as cancer. In addition, poisonous fumes could be released into the air if insulation in solar units overheated or caught fire.

Larger difficulties would arise with big solar-thermal power systems such as high-temperature thermal storage facilities, which are "like napalm," according to solar expert Henry Kelly of Congress' Office of Technology Assessment, who adds that "you wouldn't want to put them near where people lived." Nor is energy from wind environmentally perfect. Giant windmills interfere with television and radio signals and their blades can kill unwary birds. But the biggest problem is "where to put the things," Martin notes, because people in windy places like Cape Cod aren't eager to clutter their picturesque landscape with immense metal structures.[50]

Practically alone among solar enthusiasts, Ken Bossong of Citizens' Energy Project adds his voice to those pointing out the environmental problems of solar energy in a comprehensive catalog of potential ecological woes entitled *Hazards of Solar Energy.* Although brutally frank in his analysis, Bossong has the ultimate goal of promoting rather than denigrating solar energy on the practical, common-sense basis that "if solar advocates continue in their present course of promoting solar technologies as environmentally benign and safe, they will lose credibility when solar's problems begin to surface."[51]

But, if environmentalists protest the ecological problems caused by hard energy, won't they also attack soft energy when its environmental hazards begin to have an observable effect? Of course. And if they succeed, that will leave us receding to our "proper place" in the natural environment among the trees, rocks, mountains, rivers and lakes. At least that appears to be the direction of thought of Christopher D. Stone, professor of law at the University of Southern California, in a mind-bending law review article (later a book) called "Should Trees Have Standing?" Stone *seriously* proposes that legal rights be given to "forests, oceans, rivers and other so-called 'natural objects' in the environment—indeed to the natural environment as a whole."[52]

How would a "natural object" enforce its rights? Stone proposes that "special environmental legislation could be enacted along traditional guardianship lines. Such provisions could provide for guardianship both in the instance of public natural objects and also, perhaps with slightly different standards, in the instance of natural objects on 'private' land."[53] Who would be the "guardians" of these "public natural objects" and "natural objects on 'private' land" or, in effect, all of the natural environment in the country? Well, comments Stone, the "potential 'friends' that such a statutory scheme would require will hardly be lacking. The Sierra Club, Environmental Defense Fund, Friends of the Earth, Natural Resources Defense Council, and the Izaak Walton League are just some of the many groups which have manifested unflagging dedication to the environment and which are becoming increasingly capable of marshalling the requisite technical experts and lawyers."[54]

Of course. But, if natural objects had legal rights, would they also have legal responsibilities? For example, could a homeowner sue a river if it overflowed and deposited three feet of water in his basement? "Where trust funds had been established," says Stone, "they could be available for the satisfaction of judgments *against* the environment, making it bear the costs of some of the harms it imposes on other right holders." By "other right holders," Professor Stone is, let's hope, referring to human beings rather than other natural objects.

But there's just one problem, Stone notes. The business of suing a river has its difficulties because the "ontological problems would be troublesome here, however; when the Nile overflows, is it the 'responsibility' of the river? the mountains? the snow? the hydrologic cycle?"[55] One can easily imagine how a smart lawyer or even a dumb one could tie a homeowner in knots if he came into court to sue the hydrologic cycle. On the other hand, a tree and its guardian would seemingly have a much easier time of it suing a homeowner for ax damage.

Based on this brief analysis, it would appear that the natural environment would actually end up in a legal position superior to human beings if granted legal standing because it would have all kinds of rights—"I suspect that a

185

society that grew concerned enough about the environment to make it a holder of rights," exudes Stone, "would be able to find quite a number of 'rights' to have waiting for it when it got to court"—but few if any responsibilities.[56] But there's no need to worry, Stone says in effect, because the "time is already upon us when we may have to consider subordinating some human claims to those of the environment *per se* . . . we have to give up some psychic investments in our sense of separateness and specialness in the universe. . . . I do not think it too remote that we may come to regard the Earth, as some have suggested, as one organism, of which Mankind is a functional part—the mind, perhaps: different from the rest of nature, but different as a man's brain is from his lungs."[57]

But, if all of this appears too far out to be believable, it may seem even more unbelievable that Stone's proposal of providing legal standing for trees and other natural objects failed by only one vote from being accepted by the U.S. Supreme Court! This happened in *Mineral King* v. *Morton,* a 1971 case in which the Sierra Club brought suit for an injunction to prevent the U.S. Forest Service from granting a permit to Walt Disney Enterprises, Inc., which wanted to develop Mineral King Valley in California with a $35 million complex of motels, restaurants, and recreational facilities. The Sierra Club maintained that the project would adversely affect the area's esthetic and ecological balance. But the U.S. Ninth Circuit Court of Appeals ruled that the Sierra Club had no standing to bring the question to the courts because the organization did not "allege that it is 'aggrieved' or that it is 'adversely affected' within the rules of standing."[58] The Sierra Club appealed to the Supreme Court, and prior to the court's decision, Stone managed to write his "Should Trees Have

Standing?" for the *Southern California Law Review* and bring it to the attention of Supreme Court Justice William O. Douglas, himself a fervent environmentalist.

In April 1972, the Supreme Court upheld the Court of Appeals by a vote of four to three, stating that "in this case where petitioner asserted no individualized harm to itself or its members, it lacked standing to maintain the action."[60] But Justice Douglas dissented, stating that "the critical question of 'standing' would be simplified and also put neatly in focus if we fashioned a federal rule that allowed environmental issues to be litigated before federal agencies or federal courts in the name of the inanimate object about to be despoiled, defaced, or invaded by roads and bulldozers and where injury is the subject of public outrage. Contemporary public concern for protecting nature's ecological equilibrium should lead to the conferral of standing upon environmental objects to sue for their own preservation. See Stone, *Should Trees Have Standing?*"[61] Justice Blackmun also dissented, asking, "Are we to be rendered helpless to consider and evaluate allegations and challenges of this kind because of procedural limitations rooted in traditional concepts of standing? I suspect that this may be the result of today's holding."[62] Justice Brennan also dissented, agreeing that "the Sierra Club has standing."[63]

The legal proposal of standing for the natural environment thus narrowly failed to be accepted by the Supreme Court. But the concept lives on in the environmentalist movement. How the concept might be used in practice is suggested by the Sierra Club and the Natural Resources Defense Council, who have joined together in defining a "degraded environment" as any place that human actions have affected or changed. Moreover, in a 1979 policy statement, two NRDC spokesmen argued that appointed officials should fulfill the role of a trustee "as one who maintains the non-renewable environment as it was originally, to pass on to the next trustee." They also stated categorically that in view of an alleged legacy of unjust risks to unconsulted future generations, the "least unfair way of managing intertemporal relationships is for each generation to try to leave the earth as it was when they arrived."[64]

But, as Margaret N. Maxey, associate professor of bioethics at the University of Detroit, points out, such formulas as these conceal two questionable assumptions. The first is that "an untouched 'natural environment' by definition manifests a superior, if not sacred, order which human interventions violate to some degree." The second is that "a trustee of a so-called natural environment can do nothing more nor less than pass it along in its original pristine state; to do otherwise is to be guilty of some moral wrong." Overall, she observes, this "philosophy of non-degradation assumes the idea that a benign environment is rapidly being ruined by human beings. However, the historical record attests that an untamed environment has repeatedly wrought massive *human degradation* through catastrophic effects of famines, plagues, floods, earthquakes and so forth. The basic problem, therefore, is not to sustain a simplistic 'non-degradation' of the environment. Rather the problem is a complex one of devising appropriate means to protect both life-sustaining and aesthetic qualities of the biosphere, and at the same time develop technologies which provide basic human goods as a necessary condition for maintaining a preferred environmental quality."[65]

Perhaps perceiving that in goals such as standing for natural objects lies madness, the Carter administration seemingly began some belated efforts to rein in the environmental Frankenstein which it had done so much to nurture. But these efforts turned out to be less than effective. In 1978, for example, the administration established the Regulatory Council, whose mission is to monitor the economic effect of government regulations, including, presumably, the economic effect of regulations promulgated by the Environmental Protection Agency. But President Carter named as director of the Regulatory Council none other than Douglas Costle, administrator of the EPA.[66] Since environmental and consumer groups had attacked even the idea of such a council, this was undoubtedly good politics because Costle reportedly is highly regarded by these groups. But it was also clearly a case of sending the fox to guard the chicken coop.

Another try was seemingly made in 1979 when President Carter proposed an "Energy Mobilization Board" that would have sweeping powers to speed construction of as many as seventy-five "critical" energy projects at any one time. The board was to be exempt from major sections of U.S. environmental laws, and judicial reviews of its decisions were to be sharply restricted. Details of the proposal reportedly left environmentalists dismayed. "We've never seen anything like this," said David Masselli, an official of Friends of the Earth. "Its powers are half again more than we expected."[67] However, the environmentalists did not have to remain dismayed for long because within a week major administration spokesmen from the Office of Management and Budget as well as the EPA made it clear that President Carter would not do away with either the Environmental Protection Act or the Clear Air Act, even though compliance with the requirements of these acts is, in the words of Federal Energy Regulatory Administration administrative law judge George Loonis, the "real time-consuming process, the one which takes years usually. . . . The developer must prepare an impact statement, something which usually takes months, then it must be approved, again a months-long process, and then, if it is approved, the developer can almost depend on being sued by one environmental group or another."[68]

But isn't there any situation at all in which environmental restrictions might be swept aside in order to achieve needed growth in energy production? Well, yes, President Carter has the power to do so, but he would have to declare a national emergency approaching a state of actual war. In addition, Congress might give the proposed energy mobilization board power to do so. But this would only be in a "state of energy emergency," which would exist, according to an apparent general consensus of the Senate as expressed by one senator, if "we were to lose 20 percent of our crude supply through blockade, or war, or natural disaster, or what have you."[69] In other words, it's more important to protect the environment against people than it is to protect people against the catastrophic consequences of a blockade, war or natural disaster (courtesy of the environment).

12

ENERGY AND SAFETY

The growing use of energy has played a major role in enabling people to achieve socioeconomic advancement. It has also provided a human environment which is more hospitable to people's health than the natural environment. In addition, it has resulted in living conditions in which the risk of disabling or death-dealing accidents is reduced. This is because up until about a century ago people had to rely primarily on their own human energy applied to hard physical work to survive.

It is when people do physical work that accidents are more likely to happen. The growing use of nonhuman energy has made it possible to increasingly mechanize physical work that previously was done with human energy. More and more people have been able to either reduce the amount of physical work they do or switch from jobs involving physical work in heavy industries to employment in offices and stores and other service jobs requiring little if any physical effort.

The potential for reductions in accident risks resulting from this trend is shown by U.S. Bureau of Labor Statistics data which indicate that "employment in refuse collection, stevedoring, drilling and tunneling, and logging is four to five times riskier than the average for all manufacturing and thirty to fifty times riskier than clerical employment."[1] And, although there were undoubtedly other factors involved, what the increasing substitution of energy for physical work has meant in terms of accident rates is reflected to a considerable extent in statistics assembled by the National Safety Council which reports that "between 1912 and 1977 accidental deaths per 100,000 population were reduced 41 percent from 82 to 48. . . . The reduction in the overall rate during a period when the nation's population more than doubled has resulted in 1,900,000 fewer people being killed accidentally than would have been killed if the rate had not been reduced."[2]

Growing energy use therefore has resulted in increased human safety in terms of not only better health conditions but decreased danger of accident. However, this benefit of increased human safety has not been achieved without cost. This cost consists of the safety risks which are created by energy production itself. As Herbert Inhaber, associate scientific adviser of Canada's Atomic Energy Control Board puts it, every form of human activity "involves risk of accident or disease, resulting in injury or death. Generation of energy is no exception."[3] When people do physical work, they expose themselves to a host of safety risks such as muscle strains, heart attacks, self-inflicted injuries with tools, and falls, among others. So also people create safety risks when they produce energy.

But human safety has increased with energy growth, indicating that the improved safety resulting from increased energy use has more than outstripped the safety risks created in the process. As a result, we have been able to achieve the economic benefits of increasing energy consumption while simultaneously improving our safety from the standpoint of reduced risks of accidents and health problems.

The Conserver Cult, however, ignores these improvements in human safety as well as the socioeconomic advancements which have resulted from increasing energy production. It attacks conventional methods of energy production in general and nuclear power in particular as unsafe, even though both experience and numerous studies have shown that nuclear power is one of the safest energy sources of all. No member of the public, for example, has ever been injured or killed as a result of commercial nuclear powerplant operations. Furthermore, in a comprehensive study of the risks of energy production, Canada's Inhaber found that, even when the most critical estimates of nuclear safety risks in electrical generation are used, nuclear power is still 200 times less risky than coal and oil although twice as risky as natural gas.[4] In addition, the well-known Rasmussen report (WASH-1400) on reactor safety, which provided the most complete assessment ever conducted of accident risks in U.S. commercial nuclear powerplants (or any energy facilities for that matter), concluded that the risk of fatality due to a nuclear reactor accident with 100 plants in operation is one in 5 billion. Compare this to the risk of being killed in a tornado or hurricane of one in 2.5 million, or the risk of being killed in a motor vehicle accident of one in 4,000, or the risk of being killed in all types of accidents of one in 1,600.[5]

But, regardless of what experience and studies have shown concerning the safety of nuclear power, it is, of course, undeniably true that on first consideration this form of energy appears to be riskier than energy sources such as coal and oil, with which we are more familiar. Columnist William Raspberry probably expresses this as well as anyone: "I'm convinced that nuclear power plants are safer than conventional energy sources in just the same way I am convinced that air travel is safer than driving. That is, I believe it with one small rational corner of my mind. The rest of me doesn't believe it for a second."[6] However, it is also true that new forms of energy always appear riskier than the sources we are "used to," as evidenced by this quote from the February 1873 issue of *Scientific American:*

> The question of heating our buildings by steam is now about in the same condition as was the question of lighting our dwellings by [coal] gas 50 years ago. It is true that gas is the safest kind of light, much safer than kerosene or oil or candles. All the same there were scores of people raising their voices against the introduction of gas in our cities and houses. We now have exactly the same kind of alarmists among us, who, notwithstanding that heating buildings by steam is the safest of all methods— far surpassing the hot-air furnaces and infinitely safer than having a separate fire in each room—raise their voices against steam heat and attribute every fire taking place in a locality where steam is used not to the fire in the furnace but to the steam, making the pipes red hot and so igniting the woodwork in the neighborhood of these pipes.

Or consider the message which was posted in hotel rooms when they were first illuminated with electricity: "This Room Is Equipped With *Edison Electric Light.* Do not attempt to light with match. Simply turn key on wall by the door. The use of Electricity for lighting is in no way harmful to health, nor does it affect the soundness of sleep."[7]

There is nothing new about an initial public reaction of fear of an unfamiliar energy source. So there is nothing out of the ordinary about alarmists springing up among us to attack the safety of nuclear power. What *is* different is that today's attack on nuclear power safety is being waged not on the basis of objective, scientific analyses of the relative risks of all energy sources or even on the basis of simple fear of the unknown, but rather as part of an ideological, sociopolitical movement whose aim is to achieve the Conserver Cult vision of a future in which hard energy is turned off in favor of conservation and soft energy sources.

It wasn't always this way. As H. Peter Metzger relates, the antinuclear movement began in 1954 when strict atomic secrecy was relaxed. At that time, independent scientists got their first look at the government's measurements of the massive amounts of radiation that were being injected into the air by atomic-weapons testing. Through the efforts of nongovernmental scientists who challenged the claims of the Atomic Energy Commission (AEC)—which were later proven wrong—the dangers of intense airborne radiation were publicized and the Nuclear Test Ban Treaty was signed ten years later. Other dangerous activities of the AEC were also eventually revealed by outside scientists over the years, and the antinuclear movement gained strength and credibility and slowly forced additional changes in the government's plans for how the atom was to serve mankind.

Through the efforts of a few dozen scientists and lawyers, one of whom was Metzger, whose 1972 book *The Atomic Establishment* was the first serious treatment of the shortcomings of the AEC, many dangerous practices in the AEC's shaky regulatory system were tightened up. Even more important, Metzger relates, the activities of the dissident scientists and lawyers forced the government and the growing nuclear power industry to redesign nuclear powerplant safety procedures and even nuclear reactors themselves to make them even more safe. However, in 1973, after all the important conflicts had been resolved and scientists as well as many environmentalist and industry leaders were convinced nuclear power was safe, "consumer advocate" Ralph Nader announced his intention of shutting down all civilian nuclear reactors.[8]

"Nader's publicity star apparently had started to fade, and his Consumer Protection Act, not enacted to this day, was encountering serious opposition," says Metzger. "Seeking another issue, Nader turned his attention to the nuclear power industry. Late [in 1975] he staged a convention of nuclear critics in Washington, D.C., which drew 1,000 persons, and his followers then formed regional anti-nuclear groups and entered into an anti-nuclear coalition—which also is directed against western coal development—with groups like the Friends of the Earth and Environmental Action."

What would we do for energy if hard nuclear and coal resources were grounded? Conserve, of course, because, according to Nader, the fact that Sweden uses about half the energy per capita that we do but has at least as high a standard of living is "living proof that conservation does not mean a lower standard of living." In addition, Nader exulted that "if we could ever get the hobbyists of America to start building windmills and solar energy collectors out of Heathkits, [the energy savings] would be enormous."[9]

According to journalist Jacque Srouji, Nader's 1975 convention served as a "freeze frame picture of the organized nuclear opposition. . . . The anti-nuclear forces had finally crystalized under their own distinctive banner and thus blossomed into a full-fledged political movement. Interestingly enough, some of the most popular speakers were not traditional nuclear opponents but rather leaders in the anti-war and civil rights movement. Few at Nader's meeting voiced interest over distinctive issues pertaining to nuclear power (i.e., the operation of an Emergency Core Cooling System or other back-up mechanisms), but rather the assembly agreed that nuclear power was an enemy of the people and, as such, must be stopped. Their emphasis was on tactics and approaches." The convention also featured former AEC commissioner William Doub and Dr. Ralph Lapp, a nuclear consultant, who were allowed to speak at the affair but were confronted with such "When did you stop beating your wife?" questions as: "How many atomic explosions in our cities would you accept before deciding that nuclear power is not safe?—no complexities, just a number."[10]

However, in contrast to the political attack on nuclear power launched at Nader's 1975 convention, people viewing nuclear power from a strictly scientific and engineering point of view since 1975 had increasingly and overwhelmingly endorsed this new energy source. The American Nuclear Society endorsed nuclear power in 1975 in response to critics. So also did the 18,000-member Power Engineering Society. So did the Energy Committee of the 170,000-member Institute of Electrical and Electronics Engineers. So did the 69,000-member Society of Professional Engineers. So did the National Council of the 39,000-member American Institute of Chemical Engineers. So did the Board of Directors of the 3,400-member Health Physics Society." In addition, Beckmann points out that "25,000

197

scientists and engineers signed a 'Declaration of Energy Independence' urging increased use of coal and nuclear power and presented it in the White House in 1975, on the second anniversary of the Arab oil embargo. The signers of the petition had a combined total of *two hundred thousand* man-years of experience in electrical power generation."[11]

By comparison, Metzger observes, Nader and the group of antinuclear crusaders following him have "consistently publicized selected opinions of only about a dozen scientists, most of whom are zealots who espouse theories that in some cases are held by them alone." Why did this handful of scientists become nuclear critics? Metzger offers the view that "some of the older scientists . . . who once took part in the big nuclear oversell of the 1950s, might be motivated by a desire to rectify what they now perceive as their own errors. But, in the main, the middle-aged critics have been radicalized by a single bad experience, either with the old Atomic Energy Commission or with the nuclear industry from which they may have received a patronizing response to a legitimate concern."[11] As an example, he cites a *Forbes* magazine article concerning Dr. Henry Kendall, head of the antinuclear Union of Concerned Scientists: "'Take the sad case of Dr. Henry Kendall,' said *Forbes* magazine. After making his complaint, 'the AEC, in effect, told him to go jump in the lake.' Yet it was Kendall's objections that finally caused the AEC to improve its reactor safety backup systems. But Kendall 'was so soured by the experience, that he became a propagandist against nuclear development. It was Kendall, in fact, who got Ralph Nader interested in the subject.'"[12]

Both Kendall and UCS executive director Daniel Ford contend that they are "not opposed to nuclear power per se, but only to the dangerous application of it." However, Dr. Norman Rasmussen, the MIT nuclear engineer who directed the reactor safety study that has been strongly attacked by UCS, observes that "I don't think they've been reasonable in their approach. Instead of looking at each option and considering which has the most benefits and the least risks, they set some absolute [safety] standard for nuclear power that seems almost unattainable."[13]

And wouldn't setting an "almost unattainable" safety standard be as effective a way of shutting down nuclear power as an outright antinuclear stand? Not only as effective but more effective. As UCS member Jim Cubie explains in a letter, if "we define ourselves as simply 'antinuclear,' we are playing into the hands of the nuclear industry. Almost every poll indicates that the public supports nuclear energy as a general concept. However, if we define ourselves as persons who want to have nuclear safety problems solved before further commitments are made, our public support would be larger than the industry's. . . . It is essential that we attempt to change the perception of the movement from one which is simply anti-nuclear to one that is concerned about nuclear safety issues."[14]

Another example cited by Metzger is Dr. John Gofman, a professor emeritus at the University of California at Berkeley whom he describes as "the most quoted of Nader's scientist zealots." Gofman's oft-repeated charge, according to Metzger, is that the annual release of toxic and radioactive plutonium from nuclear reactors by the year 2000 will kill 500,000 persons every year, even though, as Metzger points out, the EPA estimates that the amount that would be released would be one-half ounce.

He adds that

it was not many years ago when Gofman was an active member of the same atomic establishment that, during the days of atmospheric nuclear weapons testing, dumped a total of ten tons of the stuff into the air without any objection from Gofman. Today, by Gofman's own formula, that ten tons is 100 times more plutonium than would be required to kill off the entire population of the earth each year, every year, since then. If Gofman were correct, the human race would have been wiped out long ago. Instead, the incidence of cancer has been rising at a steady rate of only about one percent a year. Since the astronomic rise in cancers to be expected before now if Gofman was right has not materialized, something must be very wrong with his calculations. I asked Gofman about his tendency to exaggerate when I visited him in his California office in 1971, shortly after he changed his mind about nuclear energy. He admitted his figures were misleading, but defended the practice by pointing out that since the atomic industry is so strong and the industry's critics so weak, such extremes were not only justifiable but absolutely necessary."[15]

Metzger describes another case involving a much publicized defection of three young General Electric nuclear engineers in California whose motivation, he points out, had

little to do with nuclear energy itself, but rather with the quasi-religious personal life style changes that can be found in virtually every area of American life. . . . [Their] reasons are diffuse and have nothing whatever to do with any special hazard or problem or example of incompetence which they might have discovered by virtue of their special position as insiders in the nuclear industry. One quit GE because of the 'uncertainty of the human factor,' another because of the American decision to sell nuclear reactors abroad, and the third because India had exploded an atomic bomb. All are members of the quasi-religious Creative Initiative Foundation which teaches that since plutonium is manmade and not God-made, it is evil. When the three eventually appeared before the Congressional Joint Committee on Atomic Energy they presented a list of things they said must be done to make nuclear power safer, but all three admitted that the problems they pointed out can be corrected.[16]

Initially, like many antinuclear critics, environmentalists were among the strongest supporters of nuclear power because it appeared to them to be a way to protect the environment. A case in point is David Brower, who was executive director of the Sierra Club until 1969, when he founded Friends of the Earth. Brower recalls that "until [1969], in fact, I was absolutely in love with the atom. . . . I was actually telling people that—by harnessing the atom—we could enter a new era of unlimited power that would do away with the need to dam our beautiful streams." However, Brower relates that he began getting a "little suspicious" in the mid-fifties when a senator on the Joint Committee of Atomic Energy told him that "you know, we're having a little problem with wastes these days." Recounting that "I didn't know what he meant then, but I know now," Brower in the seventies began saying things like "nuclear waste . . . is lethal and . . . we have absolutely no way of disposing of [it]. There is no place where we can safely store worn-out reactors or their garbage. No place!" And: "Do you *know* what a runaway reactor can do? It can produce a radioactive cloud that extends 100 miles downwind and kills everything—that could include anywhere from 10,000 to one million people—in two weeks flat. The fallout from the accident, of course, can cause major injury and damage even further away than that."[17]

The Sierra Club also took a benign view of nuclear power as a way of saving the environment from energy production, explaining that "many environmentalists have looked upon nuclear power as a potential ally. They believed that nuclear power would reduce both air pollution and the adverse consequences of stripmining because we would use less coal to generate electricity. Nuclear power would mean less offshore drilling for oil and an end to plans to dam and flood some of the world's most spectacular canyons for hydroelectric power." However, in 1974, the club's board of directors voted to oppose new nuclear power plants, stating its new policy that "the Sierra Club opposes the licensing, construction and operation of *new* nuclear reactors pending . . . resolution of the significant safety problems inherent in reactor operations, disposal of spent fuel, and possible diversion of nuclear material capable of use in weapons manufacture. . . ." Michael McCloskey, the club's executive director, added that "this policy has two sides to it. If the answers to the nuclear problems can soon be found, the industry has nothing to worry about. It merely needs to learn to speak understandably and convincingly. But if the answers do not exist, and cannot be found within a reasonable time, can we all afford to gamble so much? We are gambling with people's lives, with the well-being of future generations, and with prodigious sums of capital that may affect the viability of our economy."[18]

Many other environmentalist organizations such as the Natural Resources Defense Council today are part of the antinuclear movement which since the early seventies has produced a growing crescendo of charges, any one of which, if true, would in most cases provide sufficient reason to scratch nuclear power as a future energy option. And this increasingly appears to be the objective. Thomas J. Connolly, professor of mechanical engineering at Stanford University, for example, observes that

one very interesting social, political phenomenon is the number and the behavior of groups which have made an anti-nuclear

202

stance their principal policy position. There are variations from group to group, of course, but for the most part arguments are brought up, not for discussion, nor in an effort to achieve some solution, but simply to discredit the nuclear industry. These groups produce reports and news stories with the ease and facility of rabbits breeding, and with just about as much thought. On any given Monday one group may report that we don't need a breeder reactor program because we have such abundant uranium resources; the next day the story will be that we have so little uranium that nuclear power is uneconomic; on Wednesday another group may plant a phony story with the *Chicago Tribune* that a satellite picture has revealed that a Russian reactor has blown up; the next day it may be that a nuclear economy will deprive us of our civil rights; next will come a pseudo-scientific document purporting to show that all of the world's radiological health organizations have been low in their estimates of the toxicity of plutonium by a factor of 100,000; a group may end the week with a statement that nuclear power plants can't produce energy anyway. Now I have exaggerated the time scale to be sure, but the variety, the inconsistency, and the irresponsibility of their reports I have not. These groups operate on the principle that charges can be made much more rapidly than they can be rebutted. They know that the truth rarely catches up with false or misleading statements. They know it is easier to attack than to defend. They make their target the legitimate fears and concerns that people should have about a massive new technology. They are specialists in sowing doubt. Above all, they are experts in the use of the half-truth in situations where a half-truth is no truth at all.[19]

For example, one of the early charges made by the antinuclear movement was that nuclear powerplants could "blow up." This charge had great impact on the public and still does to this day because it seems reasonable to assume that *nuclear* bombs and *nuclear* powerplants should have an equal capacity to blow up. Thus, Helen Caldicott, an Australian pediatrician who initiated the public outcry that pressured the French government into ending the testing of nuclear bombs in the South Pacific, states that "if a nuclear war occurred, the whole of the human race would not survive. . . . Nuclear plants are synonymous with nuclear weapons."[20] And actor Robert Redford, a trustee of the Environmental Defense Fund whose home in Utah is heated with solar panels and will eventually be supplied with electricity by two windmills, declares that the "nuclear people have had their chance and now it's time to give solar an opportunity. At least it won't blow up and kill anyone."[21] However, the charge that nuclear powerplants can "blow up" is not made today by more sophisticated nuclear critics. The reason is that it is a physical impossibility. Beckmann makes this point excruciatingly clear when he states that "the idea that a nuclear plant can blow up like an atomic bomb is just preposterous. It is physically impossible. Not highly improbable, but utterly impossible. An explosive nuclear chain reaction is no more likely with the type of uranium used as powerplant fuel than it would be with pickled cucumbers. This is because uranium for explosives is enriched to more than 90% 'pure,' while uranium used to generate nuclear power is only 3.5% 'pure.'"[22]

Another charge made by critics is that nuclear power plants produce radiation that causes cancer, leukemia and genetic damage. Gofman, for example, states that

there is no such thing as a safe dose of radiation with respect to cancer, leukemia or genetic mutation injury. All authoritative bodies have held that we must operate on the basis that there will be such injuries in proportion to accumulated dose of radiation down to the lowest doses. It is not credible that the entire nuclear fuel cycle can ever contain the radionuclides perfectly, with or without accidents. Indeed such nuclides are released during so-called normal operation. Therefore, it follows that injury to humans is guaranteed the moment [a nuclear] plant starts to operate and to create the radionuclides. . . . The only way to prevent the production of the radionuclides is not to have nuclear power plants operate. . . . Since the regulatory processes do not work to protect the public, and since the regulatory authorities continue to grant licenses for the random murder of members of the public through the licensing of nuclear power plants, it is abundantly clear that the public can count upon no protection against victimization through the regulatory process.[23]

One of the first things that can be said about this statement is that it leaves the strong implication that only nuclear plants produce radionuclides or radiation. But nothing could be further from the truth. The fact is that we live in a radioactive world. Nuclear radiation is all about us because it is part of our natural environment. Radiation doses are produced not only by cosmic rays from the skies above us but by radioactive substances in the ground we walk on, the air we breathe, the food we eat, the water we drink, the houses we reside in, the buildings we work in and even our own bodies, which are mildy radioactive. As a result of all of this naturally occurring "background" radiation which has always been a part of man's environment, people in the United States are exposed to an average annual radiation dose of 135 to 150 millirems (a millirem is one-thousandth of a rem, or "roentgen-equivalent-man," the unit used to measure radiation doses to the body).

But this average annual radiation dose varies widely depending on the elevation at which we live. People living in mile-high Denver, for example, receive 75 milirems more than the national average due to a higher cosmic radiation dose. The radiation dose also varies depending on where we live because of differences in natural ground and air radioactivity throughout the country. In addition, the dose varies depending on the types of houses in which we reside. Brick houses, for example, produce 100 to 200 millirems a year; wood houses produce 25 to 50 millirems.

Variations in dosage also result from the types of work we do. Airline pilots and flight attendants, for example, receive an additional 670 millirems per year from cosmic rays, while a businessman making a round-trip, cross-country airplane trip receives five millirems on top of what he gets at home.

In addition, there are a number of manmade sources of radiation. These include medical X-rays, which expose people to an average of 50 millirems per year, fluorescent watch dials, which produce up to five millirens per year, and color television sets, which produce about 0.1 millirems per year.

The total background and manmade radiation dose received by people in the United States from all of these sources averages 150 to 200 millirems a year. But total annual doses that can be received above this average due to variations in living, working and other conditions range up to 1,000 millirems or more.[24]

The next thing that can be said about the Gofman statement is that it strongly implies that it is a known scientific fact that injuries due to radiation are directly proportional to the radiation dose down to the lowest dose level, that even a little radiation will kill you. But this is not the case. People have lived in a veritable sea of low-level radiation since the beginning of human history, during which time longevity has increased and health has improved even though additional manmade radiation has been added to the environment. If radiation were injurious even at the smallest dose, the opposite should be the case. The latency period of cancer caused by radiation is fifteen to thirty years, so the human race might possibly never have gotten out of the box.

The fact is that no one knows what the effect of the low levels of radiation occurring in the environment may be, if any. The standard assumption is that they have little or no injurious effect. It is even possible they may have a *beneficial* effect. This possibility is suggested, for instance, by the fact that Colorado has one of the highest exposures to radiation and one of the *lowest* cancer rates in the country. However, it is extremely difficult if not impossible to prove the low-level effects of radiation one way or the other. There are simply too many other variables, such such as age, socioeconomic status and occupation, which could affect the results.

What is scientifically known is that radiation is definitely injurious when it hits a level of 100 rems, which is equal to 100,000 millirems, 500 times the dose of 200 millirems produced by the natural and manmade environment. Exposure to such radiation results in three types of health effects: acute radiation sickness, cancer and genetic defects. There is substantial data indicating the incidence of acute radiation sickness and cancer in human beings as a result of high radiation doses. But this is not the case with respect to genetic defects, which have been observed only in animals, not humans. Studies of the 350,000 survivors of the Hiroshima and Nagasaki atom bombs, for example, have yielded no evidence of genetic damage among offspring due to radiation. As a matter of fact, the 350,000 Hiroshima and Nagasaki survivors are doing better than expected. According to Prof. Joseph Rotblat, who worked on nuclear bombs at Los Alamos after the war, they have more leukemia and other forms of cancer than the rest of the Japanese population, but nonetheless they have a lower overall mortality rate.

Furthermore, no genetic effects have been found in the children of the survivors, nor have there been increases in cancers in people who were in their mothers' wombs when the bombs hit. Contradicting data gathered from animal experiments, these findings could just as well be used to suggest that "a little radiation is good for you," just as a little water is not only good but absolutely necessary for health while a lot of water taken all at once can result in drowning.[25]

But what really can be said from a scientific viewpoint about the health effects of low background radiation levels of 100 to 200 millirems, one five-hundreth or less of the dose of 100 rems at which significant health effects have been observed in human beings? Nothing. No scientific data is available to show what these effects may be, although common sense suggests there is little or no effect. However, to be on the safe side, scientific organizations responsible for setting radiation exposure standards take the extremely cautious and conservative approach that something *can* be said, even though data is lacking, by *assuming* that there is a proportional or linear relationship between radiation dose and injury and that there is no radiation level or threshold below which radiation has no effect. As has been pointed out by Dr. Bernard L. Cohen, professor of physics and director of the Nuclear Physics Laboratory at the University of Pittsburgh, it is "usually assumed that the cancer risk is proportional to the total exposure in rem. This 'linear-no-threshold' hypothesis involves a very large extrapolation; it assumes, for example, that the probability of cancer induction by one millirem, which is typical of most exposures of interest, is 10^{-5} times [100,000th of] the probability of induction by 100 rem, which is the region for which most data are available. For a number of reasons, however, this assumption seems more likely to overestimate than underestimate the effects of low dosage." Cohen goes on to say that all four major national and international scientific organizations reponsible for monitoring radiation have acknowledged that the "linear-no-threshold" hypothesis is highly conservative, and although they all accept it as a basis for setting exposure standards, only one condones it for estimating risks, while two others either pointedly refuse to do so or are highly critical of it.[26]

Finally, Gofman explicitly says that regulatory authorities "grant licenses for the random murder of members of the public through the licensing of nuclear power plants." This implies that nuclear powerplants produce huge amounts of radiation sufficient to kill people. But how much radiation is each person in the United States actually exposed to by nuclear powerplant operations each year? Not 100 rems or 10 rems or even 1 rem. The exposure is less than one millirem, more than 100,000 times *less* than the 100 rems at which significant health effects have been observed. Furthermore, it is estimated that, assuming nuclear energy becomes a dominant source of electricity by 2000, each person in the United States will still receive an average yearly dose of less than one millirem, while people living near nuclear powerplants will receive no more than five millirems a year.[27] Five millirems is 40 times less than the high-average, existing background radiation of 200 millirems a year which, in turn, is 500 times less than the radiation dose of 100 rems at which significant health effects can be observed. All of which suggests that the murder found in claims that nuclear powerplants kill people is not of people but of scientific fact.

But what if a nuclear power plant has an accident? An accident would release huge amounts of radiation which would kill thousands, if not tens of thousands, of people, wouldn't it? This is the charge made by nuclear power critics such as Daniel Ford of the Union of Concerned Scientists, who states that "nuclear plants of current design are susceptible to catastrophic nuclear radiation accidents because their safety was compromised in the effort to move forward quickly with nuclear powerplant construction. .... In unfortunate circumstances, tens of thousands of fatalities could result both from acute nuclear radiation injuries and from long-term effects among the exposed population."[28] However, if the safety of nuclear powerplants has been "compromised" as Ford states, one would expect to find a record of injuries and deaths resulting from plant operations, just as a statement that the safety of automobiles or power mowers or electric razors had been "compromised" would lead one to look for a trail of car crashes or mutilated limbs or nicked chins. But no such record of injuries or deaths resulting from nuclear powerplant accidents exists. Just the opposite is the case. In the more than twenty years of operation of commercial nuclear powerplants, no member of the public has even been injured much less killed. This is a record which one might well wish other producers would strive to achieve by similarly "compromising" the safety of their products.

The reason for this excellent record is, of course, that safety has always been of paramount concern to the nuclear power industry as well as to government regulatory authorities. The primary safety problem which must be guarded against in a nuclear powerplant is the possibility of a "meltdown," an event which has also never occurred in more than twenty years of nuclear operations. Resulting from an extended loss-of-coolant accident, or LOCA, which would leave a nuclear reactor's fuel core exposed above the level of its cooling water, a meltdown would result in the release of massive amounts of radiation. To guard against a meltdown, nuclear power-plants are designed with what is called "defense in depth." The first level of this defense requires that plants operate with a high degree of assurance that no accidents will occur that might lead to a LOCA and meltdown. But, since nothing is perfect and some accidents can be expected to occur regardless of how good defenses are at this first level, a second level is developed to detect and either arrest or safely accommodate accidents at the first level. The defense-in-depth then goes one step further by developing a third level, which is designed to provide protection against the possibility of a failure in the second-level defense at the same time that an accident it is designed to control occurs in the first defense-level. One of the third-level defense systems used in all nuclear powerplants is a redundant emergency core cooling system, which is used to automatically cool the fuel core in the event of failures of normal and back-up cooling systems in the first and second defense-levels. Another third-level defense is the containment building, which is designed to prevent radiation from leaving a nuclear powerplant *even if all other systems fail* and a meltdown occurs.

All of these first-, second- and third-level safety systems in a nuclear powerplant are designed to defend against all anticipated accidents. But, at the same time, the final overall safety objective is to provide protection even in the case of an unanticipated accident, in other words, *no matter what happens.* It is for this reason, for example, that the containment building is made of four-foot-thick, heavily reinforced and steel-lined concrete, which makes it capable of not only containing any radiation that may be released but also of withstanding hurricanes, tornadoes, earthquakes and jumbo-jet crashes. As Professor Beckmann puts it, the "material and structure of the containment building is not unlike the U-boat pens built by the Germans on the French coast during World War II. In spite of savage round-the-clock bombing and the use of special 'blockbuster' bombs, the Allies failed to crack them"[29]

Why go to all the trouble of this defense in depth? Because it was known from the beginning of nuclear power development that without these safety features an accident could result in serious loss of life and damage to property. To determine just how much loss of life and damage to property would result from a major nuclear accident was the purpose of a report (WASH-740) produced for the Atomic Energy Commission in 1957. Preparation of this report presented a problem, according to Dr. Herbert J. Kouts of Brookhaven National Laboratory. The problem was that to

produce a major accident, it must be assumed that somehow an accident has started and that no safety features have been effective at all. Fission products must be assumed to have escaped in succession each of the barriers normally confining them. . . . Furthermore, a number of additional safety systems required for every modern nuclear reactor would also have to fail. . . . The list of successive protective features is so impressive that the study group concluded that the probability of a large accident that would submit the public to hazards from fission products is incredibly small. . . . In spite of this conclusion it was still necessary to provide . . . estimates of the magnitude of effects of the hypothetical incredible accident. An analysis was therefore made of the consequences of accidents in which no safety features were assumed to function. In fact, the analysis completely ignored the existence of these features, and this was equivalent to assuming that *no safety features had been present in the first place* [emphasis added]. No realistic method of dispersing the fission products about the environment was known. It was therefore assumed that by some unknown process an instantaneous dispersion of the power reactor took place. Calculations were then made of the distribution and deposition of fission products as a function of time. . . . For the worst combination of meteorology and wind speed, taking into account no safety features of the reactor, and assuming instantaneous release of the fission products, estimates ranged to several thousand deaths and several billions of dollars of damage requiring compensation.[30]

Although it was published in 1957, this study of a "hypothetical incredible accident" based on the assumption of a *complete absence of safety features* as well as other incredible assumptions apparently became the basis twenty-two years later for a famous line in the 1979 movie *China Syndrome.* (Someone once joked that if a meltdown occurred, the resulting molten ball of nuclear fuel would burn a hole in the ground all the way to China. Ergo, the "China Syndrome.") All who saw the film will recall the line that caused audiences to gasp not only because of the enormity of the catastrophe which it predicted but the remarkable coincidence between the line and the accident that had recently occurred at the Three Mile Island nuclear powerplant near Harrisburg, Pennsylvania. Uttered by a supposed expert viewing a film of a nuclear powerplant control room during a potential LOCA, the line went as follows: "If the core is exposed, [it could] render an area the size of Pennsylvania permanently uninhabitable." According to Aaron Latham in *Esquire,* when screenplay writer Michael Gray "added the near meltdown to the script, he also added the line about the possibility of such an accident depopulating an area the size of Pennsylvania—*which was taken from an actual study* [emphasis added]."[31]

213

The study the line was taken from was in all probability a 1964–65 update to WASH-740, in which there appears in a December 16, 1964, memo on the minutes of the steering committee the statement that "for a big accident the area [of disaster] would be the size of the State of Pennsylvania."[32] In the WASH-740 this line made sense because the report described a situation *in which safety systems were completely absent while weather and other conditions conducive to a major disaster were perfect.* But in *China Syndrome,* which supposedly was presenting an authentic, realistic story of a possible accident in a nuclear powerplant, the line can only be described as totally misleading. Perhaps a sequel will explain how all the safety systems that are an integral part of modern nuclear powerplants were to be whisked out of existence in order to "render an area the size of Pennsylvania permanently uninhabitable." For, in the real world, the core at the Three Mile Island nuclear powerplant was, in fact, exposed in what turned out to be the most serious accident to date in a commercial nuclear powerplant. But, although the accident caused great public anxiety and an immense financial loss, it also provided a textbook example of how effectively a nuclear powerplant's defense-in-depth works in actual practice.

The accident was precipitated by a Rube Goldberg series of unlikely events, including several equipment malfunctions, two pumps that were shut off when they weren't supposed to be, and human interventions that, although perhaps seemingly justified at the time by the information available, still resulted in exacerbating rather than alleviating problems at the plant. In fact, the Nuclear Regulatory Commission reported following the accident that it might have been prevented "if plant operators had let safety equipment function as it was designed to do."[33]

Due to operating problems, a small LOCA, or "loss-of-coolant," accident occurred. Bringing up the third level of defense, the plant's emergency core cooling system immediately and automatically came on stream. However, plant operators turned the emergency core cooling system off for reasons which later were found to be mistaken, resulting in exposure of the core and release of radiation. The radiation was prevented from reaching the outside by the containment building. Meanwhile, the emergency core cooling system was brought back on line when plant operators realized that turning it off had been a mistake. What little radiation was emitted to the atmosphere came from an auxiliary building into which radioactive water was mistakenly pumped. But even this radiation was filtered before it was released. When the radioactive water was pumped back into the containment building, some additional "puffs" of radiation were released due to spillage. But these were also filtered. As a result, actual radiation emitted to the outside was reduced to a bare minimum.

So the dread accident in *China Syndrome* which supposedly was going to render Pennsylvania "uninhabitable"—a fuel-core exposure—actually happened at the Three Mile Island nuclear powerplant, in Pennsylvania.

Who was killed by the accident? Nobody.

Who was injured by the accident? Nobody.

What was the maximum radiation dose that an individual outside the plant might have received as a result of the accident? About 100 millirems, the equivalent of two medical X-rays. But this would have required that the person stand at a point one-half mile from the plant twenty-four hours a day for a ten-day period following the accident, according to the Bureau of Radiological Health which studied data collected at the site.[34]

215

What was the average radiation dose received by an individual outside the plant? One and a half millirems, less than one-third the maximum radiation dose produced by a fluorescent watch dial.[35]

What is the projected number of excess fatal cancers due to the accident that could occur over the remaining lifetime of the population within 50 miles of the Three Mile Island nuclear powerplant? One. But this "one" will be hard to pinpoint because, if the accident had not occurred, the number of fatal cancers that would normally be expected in a population of this size over its remaining lifetime is estimated to be 325,000.[36] It should also be noted that the estimate of one cancer death comes from risk estimates based on the assumption of a "linear-no-threshold" relationship between radiation doses and health effects, an assumption which is intentionally on the cautious and conservative side. In addition, the estimate of one cancer death is derived from several other "simplifying" assumptions which "ignore factors that are known to reduce exposure" and "introduce significant over-estimates of actual doses" in order to "ensure that the estimate erred on the high side."[37] For example, no reduction in estimates was made to account for shielding by buildings when people remained indoors or for people who simply left the area. Furthermore, no reduction was made to account for the fact that the actual radiation dose absorbed by internal body organs is less than the assumed dose.

What, then, in summary, can be said about the health effects of the accident? It does not appear that there is any risk of radiation damage even to babies in their mothers' wombs, the class of people who are most susceptible of all to radiation damage because of the fast growth rate of their cells. This was the conclusion arrived at in a joint statement by the American College of Obstetricians and Gynecologists and the American College of Radiology.[38]

But what if there had been a meltdown? The Three Mile Island nuclear plant never came anywhere near a meltdown. A meltdown requires a temperature of 5,000 degrees Fahrenheit. The temperature in the reactor never got higher than about 2,000 degrees Fahrenheit. In addition, as Boston University's Samuel McCracken points out, Three Mile Island makes meltdown seem *"less* likely than we had thought. It had been widely assumed that reactor cores could not be uncovered for more than a few minutes without starting a meltdown . . . yet the Three Mile Island core was uncovered for a great deal longer than a few minutes . . . estimates range from two to 15 hours."[39] And, remember, the containment building was there to prevent release of radiation if there had been a meltdown—which again, never came close to happening.

Yes, but what if that hydrogen bubble in the reactor had exploded, then what? The hydrogen bubble made three-inch headlines on front pages of newspapers all over the country. But there was never any chance that the bubble would explode. Hydrogen needs oxygen to burn. There was no free oxygen in the reactor, so there was no way the hydrogen bubble could have exploded. This was acknowledged by Roger Mattson, systems safety division director for the Nuclear Regulatory Commission, when he said after the incident that "we fouled up. . . . There never was any danger of a hydrogen explosion in the bubble. It was a regrettable error."[40] Of course, this statement wound up in the truss ads.

Okay, but didn't the experts say that an accident like this couldn't happen? No. As Edward Teller comments, experts such as those who prepared the Rasmussen report estimated that one small LOCA such as occurred at Three

Mile Island should be expected every "100 reactor years if one is pessimistic, or in 10,000 reactor years if one is optimistic. It [Three Mile Island] occurred a little less than 500 reactor years if all the reactors in the United States are counted, and a little more than 1,000 reactor years if we count all the electric generator reactors throughout the world . . . In the end, when all information is in, we will have learned more about nuclear generation, and reactors will become even more safe. This knowledge will have caused the loss of at least $100 million, possibly a lot more. It will have been bought at no cost to human life or health. This is as it should be."[41]

But perhaps the best perspective of the accident at the Three Mile Island nuclear plant came from abroad. Here's what England's William Waldegrave, a member of Parliament and secretary of the Conservative backbench energy committee had to say: "I recently read in the *Guardian* a reference to the accident at Harrisburg nuclear station in America. It was described as 'the Harrisburg disaster.' Well. If that was a disaster, let's have more disasters like it; no one was killed or hurt, and there were, so far as I know, no long term damaging effects. On the contrary, if you want to test a reactor system to the limit, and far beyond it, by what appear in retrospect as a series of amazing operating blunders, what happened on Three Mile Island shows how tolerant nuclear equipment is. I would be prepared to bet that a big chemical plant, treated in that way, would have killed a lot of people."[42]

Does all of this mean that a catastrophic accident cannot occur in a nuclear power plant? No. The Rasmussen report, for example, states that with 100 nuclear plants in operation accidents involving 100 or even 1,000 or more fatalities could occur. However, the report also states that the probabilities of such catastrophes occurring are respectively one in 100,000 years and one in 1,000,000 years, the same range of probabilities that like numbers of people will be killed by a meteorite.[43] A Risk Assessment Review Group chartered by the Nuclear Regulatory Commission to review the Rasmussen study made the primary criticism that the actual ranges of high- and low-risk were probably greater than reported in the study. But the group added that "despite its shortcomings, WASH-1400 provides at this time the most complete single picture of accident probabilities associated with nuclear reactors."[44]

Antinuclear activists have also criticized the Rasmussen report, claiming that it understates the average number of fatalities from reactor accidents by a factor of sixty. But, even if this is the case and all other possible fatalities due to nuclear power in an all-nuclear economy are taken into account, the risks of nuclear power are still extremely small. Bernard Cohen points out that they would be equivalent to the risks of smoking three cigarettes per year, or spending twenty days of one's life or one day every three years in a city instead of a rural area, or riding in automobiles an extra hundred miles a year, or riding a hundred miles a year in a small instead of a large car.[45]

However, just as nuclear powerplants are designed with a defense-in-depth, antinuclear critics operate with what might be called an "attack-by-avalanche," in which no aspect of nuclear power escapes criticism. Critics, for example, maintain that wastes produced by nuclear plants are not only unacceptably hazardous but technologically impossible to handle. But nuclear wastes are first of all extremely small. If U.S. electric power capacity were completely nuclear, for example, the total volume of all wastes per person per year would be the size of an aspirin tablet. Because of this small size, the problem of permanently disposing of nuclear wastes is not particularly urgent; they can be safely kept in temporary storage for decades if desired. However, methods have been developed to provide permanent storage for nuclear wastes. Basically, the wastes would be sealed in highly durable glass or ceramic, enclosed in stainless steel cylinders and buried deep underground in salt or rock formations which have been stable for millenia. The wastes, it is true, would remain radioactive for thousands of years as nuclear critics hysterically cry. For example, the halflife of plutonium 239 is 24,400 years, but this figure is both meaningless and misleading because within 600 years the wastes will naturally decay to a point where they are no more radioactively dangerous than the original uranium ore mined to make the fuel from which the wastes were produced in the first place. And the wastes will be buried thousands of feet below the surface of the earth in sealed containers, whereas uranium ore is dispersed at random in its natural state. It is for all of these reasons and more that countless studies and reports, including more than 5,600 by federal government agencies alone, have resulted in an overwhelming consensus that radioactive waste can be handled and disposed of safely, with no serious public health and safety or environmental effects.

To take just one of many possible examples, the "Report to the American Physical Society by the Study Group on Nuclear Fuel Cycles and Waste Management" states that "for all LWR [light water reactor] fuel cycle options, safe and reliable management of nuclear waste and control of radioactive effluents can be accomplished with technologies that either exist or involve straightforward extension of existing capabilities. However, technical choices, including those for geologic waste disposal, require further delineation of regulatory policies. For normal operation of all fuel cycle options studied, potential radiation exposures from either wastes or effluents do not appear to limit deployment of nuclear power."[46] In other words, what is lacking in waste management is not technological capabilities but political and regulatory decisionmaking.

But it would be extremely shortsighted as well as the antithesis of good conservation practice to simply bury nuclear wastes in safe geologic formations for eons to come. These "wastes" can be reprocessed to extract plutonium, which can be used to fuel reactors, thus vastly expanding the potential of nuclear power. However, antinuclear critics go into a positive frenzy when plutonium is mentioned, claiming, among other things, that: (1) plutonium is the most toxic substance known to man, and (2) "a pound of plutonium can cause eight billion lung cancers" (Ralph Nader), and (3) "a single particle of plutonium inhaled into the lung can cause cancer" (Senator Abraham Ribicoff). But, as Bernard Cohen points out, there is "nothing in the scientific literature to support these statements which are completely contradictory to the thinking of the scientific community in radiation research." To prove his point, Cohen offered to go on television and: (1) eat 800 milligrams of plutonium 239 in its usual oxide form or an amount of it equal to as much caffeine as any prominent critic would be willing to eat (Cohen's research showed caffeine to be about as toxic as plutonium) or (2) eat five times as much plutonium as the quantity of potassium cyanide or mercury dichloride eaten by any critic (his research showed these substances to be far more toxic than plutonium). He also offered to inhale 1,000 times the quantity of plutonium that Nader implied was lethal, in any form Nader specified, and inhale 1,000 particles of plutonium of any size that could be suspended in air for a few minutes. Summing up, Cohen stated that "these plutonium intakes would increase my risk of cancer by an amount equivalent to that of moving from the Southeastern U.S. to New England, or from a rural area into a city (where I now live). I consider this risk to be well worth the benefit achieved by straightening out the thinking of the American people on this vital issue." However, he had no takers.[47]

Another charge made by antinuclear critics is that commercial nuclear power operations create a risk of nuclear proliferation because of the possibility that the plutonium produced in the reactors will be used to make atomic bombs. But, as Beckmann points out, there are "eight well-known ways of producing nuclear explosives (and three more are on the way); of these, extraction of plutonium from spent fuel rods (produced as a result of nuclear operations) is the most time-consuming, needs the most expertise, and costs about ten times more than the next most difficult method. None of the other methods even remotely involve nuclear power."[48] In addition, a new type of breeder reactor has been developed which will not produce plutonium pure enough to be used in bombs, effectively rendering the problem null and void.[49]

"But," the antinuclear critics cry, "terrorists will steal the plutonium and fashion homemade bombs with which to blackmail whole populations!" As Beckmann puts it, "the idea of a *domestic* band of terrorists obtaining enough plutonium to fashion a crude bomb from it makes good material for TV horror stories (masquerading as 'documentaries') and high school science circle essays, but the implementation in reality borders on the absurd. One might list the long series of natural obstacles even in the absence of safeguards, and one might then list the long series of easily implemented safeguards . . . but the idea is too absurd to deserve much space in a short essay. Instead, let us remember that it would be easier (not easy, but easier) to divert a ready-made weapon, and that only a supremely incompetent band of terrorists would attempt the project of a home-made bomb—dangerous, time-consuming, and almost certainly doomed to failure—when there are so many more deadly and far more easy ways of inflicting, or even threatening to inflict, loss of life in the hundreds of thousands."[50] However, it should be noted that when antinuclear critics continually claim that terrorists *could* make their own nuclear bomb, they are in effect *preparing* the public to accept this possibility as realistic, thus increasing not only the terrorist incentive to make such claims but the likelihood that their blackmailing bluffs will be believed.

The list of antinuclear charges most certainly does not stop here. But all of them partake of what Samuel McCracken calls a "common habit among the anti-nuclear lobby: holding nuclear reactors to a standard of safety that if generalized would forbid not only all other forms of power generation, but also most of the things man makes or finds in nature."[51] Or, as Scientists and Engineers for Secure Energy expresses it, antinuclear arguments are supported by "quotations taken out of context, untested assumptions, and emotional rhetoric; scientific data have been stretched far beyond their area of reliability. In defense of their pre-conceived stands, some participants in the debate have even raised improbable scenarios of the future, which can lead only to confusion and irrational fear."

The organization of scientists and engineers goes on to say that

> nuclear energy is safe. This does not mean it entails no risk whatever; every aspect of human existence carries some degree of risk. Whether we sit at home, or walk a street, work in an office or a factory, ride a car or swim a lake, there is always the possibility of a mishap. The prudent person must therefore be willing to compare risks. Consequently, in any discussion of nuclear power, it is vital to keep in mind the fact that the risk associated with the production and use of such power is in fact *less than or comparable to* the risks associated with other aspects of our daily life. If this risk were sufficient to justify banning all nuclear power production, one could then justify the absurd banning of a vast number of other human activities. . . . The average citizen is in greater danger of being injured by debris falling from a high-rise building or choking on a morsel of food. The emotional level of the nuclear energy debate tells us something about the degree of misunderstanding and misapplication of the concept of risk.[52]

Furthermore, nuclear power is not only as safe as or safer than a "vast number of other human activities," and safer than hard energy sources such as coal and oil, it is also safer than all forms of soft energy with which the Conserver Cult would replace it because it is supposedly unsafe. Here, for example, are the figures developed by Herbert Inhaber for all major sources of energy, including soft energy, in terms of public and occupational man-days lost per unit of energy output (according to Inhaber, it was not possible to draw firm conclusions concerning hydropower due to lack of data):[53]

As the table shows, nuclear power has the lowest risk of any energy source but natural gas. Soft energy sources have much higher risks, and oil and coal are riskier still. What are the reasons for these results? "Conventional

Energy Source	Public	Occupational	Total
Natural gas	—	5.9	5.9
Nuclear	1.4	8.7	10.1
Ocean Thermal	1.4	30.0	31.4
Solar Space-heating	9.5	103.0	112.5
Solar Thermal	510.0	101.0	611.0
Solar Photo-voltaic	511.0	188.0	699.0
Wind	539.0	282.0	821.0
Methanol	0.4	1,270.0	1,270.4
Oil	1,920.0	18.0	1,938.0
Coal	2,010.0	73.0	2,083.0

energy sources like coal, oil, natural gas and nuclear have lower material and construction requirements than non-conventional sources," Inhaber explains. "In some cases, they are more than 300 times lower. This is reflected in the occupational risk: the more material used, the greater this risk. . . . Methanol has the highest value, with over 200 man-days lost due to the dangers of cutting and transporting wood. But, in terms of public risk, the ranking is somewhat different, Inhaber points out. Natural gas is again lowest. However, due to its lack of air pollution, methanol is second. Nuclear power and ocean thermal have the next lowest public risk values. Wind, solar thermal and solar photovoltaic each have substantial public risk, mainly due to air pollution caused by smelting steel, a required metal for these technologies. Finally, coal and oil have the highest public risk, again due to pollution effects.[54]

Inhaber arrived at a low public risk value for nuclear power despite his following a conservative approach of taking estimates of its accident risk from a well-known opponent of nuclear power, Dr. John Holdren, professor of energy and resources at the University of California, Berkeley. Holdren had written several articles critical of nuclear power, including one coauthored with antipeople, antienergy activist Paul Ehrlich.[55]

Following publication of Inhaber's report, Holdren launched an attack on it which was described by *Wall Street Journal* editorial staff member Suzanne Weaver as "egregious in its invective and threats." (Ehrlich meanwhile reportedly began planning an article on what he called the "Inhaber Report Scandal" for *Mother Earth News*.[56]) In a letter to the *Journal* alluding to Weaver's comments, Holdren stated that "I have not threatened anyone with anything," adding that "as to the directness and firmness of my language in this controversy, I have no apology to make. By the time I entered the fray, Dr. Inhaber and the Atomic Energy Control Board had already demonstrated their unresponsiveness to correctives applied with the usual professional niceties. . . ." In the same letter, Holdren stated among other criticisms that "Inhaber's 'current' risk estimates . . . have not in fact corrected the biggest errors. For example, the 'upper limit' of the nuclear risk shown still does not encompass the upper limits given in the work cited by Inhaber (including my own) as the sources of these data."[57]

Inhaber responded:

Prof. Holdren states that I have made substantial errors. The major one he specifies is that I have greatly under-estimated the risk of nuclear power. The value of total (occupational and public) risk I assumed for this energy source was about twice that estimated in a long report Prof. Holdren himself wrote (with two graduate students) in 1975. In terms of public risk, due to accidents at reactors, the values I assumed were between 17 and

226

40 times what Prof. Holdren estimated in the summarizing tables of his report. I do not see how it can be said that I have underestimated a quantity when I have used much larger values than Prof. Holdren himself estimated. In spite of Prof. Holdren's tone of voice in his comments, I have carefully checked each of his statements. I find that those with merit do not substantially change, in terms of risk, the relative rankings of the energy systems I considered.[58]

However, although nuclear power is safer than not only coal and oil but soft energy sources as well, the Carter administration has effectively gutted its development at the behest of antinuclear Conserver Cultists. The Natural Resources Defense Council, for example, tells how "for the last six years NRDC has taken a leading position in opposing a national commitment to the 'plutonium economy.' In April, 1977, these efforts influenced a reversal of United States policy—a decision by President Carter to abandon indefinitely nuclear reprocessing and development of plutonium breeder reactors."[59] Then-Energy Secretary Schlesinger in an interview with *Time* admitted that the decision to defer recycling and the breeder was "largely a concession to the environmentalists."[60] Only congressional support of the breeder reactor has kept it alive. Meanwhile, Gustave Speth, former NRDC attorney in his new role as a Carter-appointed member of the Council on Environmental Quality called for the abandonment of nuclear power if the problem of its "deadly wastes" could not be resolved by a certain deadline while recommending that "no new electric generating units be added to the nation's grids until it is 'absolutely certain' that there is no solar or conservation substitute."[61]

In addition, nuclear power development the administration has not stopped by fiat has just as effectively been brought to a halt through bungling, which to a large extent has been caused by attempts to make further concessions to never-ending Conserver Cult demands. Metzger tells how in 1977 "nuclear licensing was suspended in order to streamline the process and shorten the twelve-year-long licensing period. About halfway through this period, the *New York Times,* a not-exactly-pro-nuclear newspaper, said in an editorial: 'The Carter planners have operated in such a hasty and even slapdash way that their draft legislation is itself now in need of study and thorough review.' The *Times* went on: 'Citizen [read: Conserver Cult] efforts to point out potential problems would be subsidized, but the administration's proposals are so loosely worded that even skilled observers can't predict whether they will strengthen or weaken present licensing safeguards.'" Metzer adds that the Nuclear Licensing Reform Bill was submitted to Congress in March 1978. But "its future looks dim as anti-nuclear zealots both in and out of government zero in on legislation they call 'unacceptable' (that's the opinion of Nader's Congress Watch) and 'badly conceived, badly drafted, badly motivated and a sell-out' (the opinion of the Natural Resources Defense Council). As if that wasn't enough, there are 'public interest' lawsuits today, by the score, designed to stop nuclear power. They range all the way from overt efforts to shut down all nuclear reactors immediately to the sneaky kind which, wolf-in-sheep's-clothing style, merely attempt to cripple nuclear economics so badly that nuclear becomes an unattractive investment for a utility."[62]

Senator James A. McClure describes the results of the Carter Administration policies as a

sad reality. A reality that while the rest of the industrialized world charges ahead with their own nuclear power technology, the United States abdicates its leadership in the development of nuclear power. . . . Internationally the Administration attempted to lead other industrialized nations down his [Carter's] self-denial path, seeking curtailment of their breeder development programs. The result, as you might well expect, is that while the United States delays its own breeder program, the rest of the world is proceeding towards demonstrations necessary for their own domestic programs. The French are now constructing their third demonstration breeder, and we have yet to begin our first. . . . The Administration policy on nuclear power development in the United States has all the trappings of a giant marshmallow. The President in his April 1977, National Energy Plan announced that the United States would use nuclear power only as a 'last resort'. . . . Since that time, we have had a number of emergencies, including the winter of 1976/77 natural gas shortage, the winter of 1978 coal strike, and the current interruption of Iranian oil exports. During this time, this very same Administration has encouraged, and at times begged, domestic utilities to run their nuclear plants at maximum power to provide the required substitute power for lost electrical generation. . . . Yet, steadfastly, the President has refused for the past two years to reconsider or modify his blatantly antinuclear stand.[63]

Following the Three Mile Island nondisaster, the antinuclear movement held a rally in which a reported 65,000 demonstrators marched on the Capitol in Washington to "shout their opposition to atomic-powered electricity" and to "roar their approval" when asked to "join in the politics of the future" by California Governor Jerry Brown.[64] President Carter served up a typical "marshmallow" of the type described by McClure when he told protest leaders that "it's out of the question to shut down all the nuclear powerplants in this country," adding in the same breath that "we do, however, want to shift toward alternate energy supplies and also strict conservation programs to minimize the requirement for nuclear power."[65] But Ralph Nader, who had previously called nuclear energy "our technological Vietnam,"[66] shouted to the crowd that "the use of nuclear power is too unsafe, too uneconomic, too unreliable to be tolerated in our land! Can't we find a better way to boil water?" Environmentalist Barry Commoner told the crowd that "today is the dawn of a new age in the nation's history, the day the solar age was born and the nuclear age died." Even Jane "China Syndrome" Fonda got into the act, proclaiming that "putting James Schlesinger in charge of solar power is like putting Dracula in charge of a blood bank." And Senator Kennedy sent a message that proclaimed: "Unless we can build [nuclear plants] safely, they should not be built at all." Almost all the demonstrators, it was reported, were white, young and the kind of upper-middle-class people who dress casually for effect rather than for economy. "We could have used fewer faded jeans and more checked double-knits," one of the organizers said.[67]

However, what was not mentioned at the rally—some things, of course, are simply unmentionable—was what it would mean to the American people if nuclear power development were slowed down or stopped by a moratorium or if all nuclear powerplants were completely abandoned. Senator McClure points out that, as far as the existing slowdown in nuclear power development is concerned, what is "most distressing, most potentially disastrous for this nation is the loss of time. Time which is needed for research and development; time which is absolutely essential to plan, design, and construct nuclear generation facilities to say nothing of the interminable delays in licensing requirements. For it is time which is rapidly closing in on us, which will cause us to suffer energy shortages in the near future—shortages which are as predictable as the sun rising in the East tomorrow morning."[68] Time, in fact, is the only truly unrenewable resource.

According to Walter D. Meyer, Myron Rollins and Raymond W. Williams of the nuclear engineering department at the University of Missouri, a nuclear power moratorium would result in "additional costs to the consumer with no compensating improvement in the quality of life. In fact, the necessity to burn additional coal and oil to make up for the loss of nuclear generating capacity will hasten deterioration of both the environment and the economy of the U.S. with an inevitable deterioration in the quality of life."[69]

What would happen if, instead of a moratorium, nuclear powerplants throughout the nation were simply abandoned, period? This was the question that the Nuclear Regulatory Commission had to answer in response to a petition from a lady in Nashville, Tennessee, alleging that nuclear energy is producing fatalities among innocent members of the public and that NRC licensing actions represent "crimes against humanity of the most heinous degree." Here is what the NRC found: The immediate economic penalty associated with abandoning nuclear power would be $60 billion to replace currently licensed reactors and higher electricity costs reaching $9.8 billion a year over the next decade. Over the long term, the impact on industry and commerce could be "measured in tens to hundreds of billions of dollars per year depending on how successful generating plants burning other fuels are at replacing the nuclear deficit."[70] Quite a price to pay for what would amount to a reduced level of safety in energy production.

13

ENERGY AND PRICE CONTROLS

Energy growth has resulted in better living standards, improving health and environmental conditions and increasing safety. But these benefits of growing energy use could not have been achieved if there had not been a long-term decline in the price of energy, making expanding consumption economically possible.

The price of energy has declined over the years as a result of the growing use of increasingly more economical sources of energy in place of human energy, the most expensive energy source of all. New, more economical energy sources have been developed by energy producers in response to increases in prices of existing energy sources. For example, wood was a major source of energy in the nineteenth century. Initially, wood was plentiful and its price was low. However, with continuing growth in use, wood became scarcer and its price went up. This created a financial incentive for producers to replace wood with a more economical energy source. This turned out to be coal, which, as a result of lower prices, displaced wood over time. The American "whale oil crisis" of the 1800s discussed in chapter six is another example of how increasing energy prices spur the development of lower-priced substitutes.

Declining energy prices came about not as a result of government price controls but solely due to the automatic "price controls" provided by competition among energy producers in a free market. Consider oil. As D. T. Armentano, professor of economics at the University of Hartford, puts it, the

> early years of petroleum industry development (1859–1911) are remarkable in that they represent a near textbook example of a *laissez-faire* market economy. There was little government regulation or subsidization during this period (no price controls, entry restrictions or tariffs) and, not coincidentally, the industry experienced a phenomenal growth and development of resources. Outputs of kerosene and related products were enormously expanded and prices were reduced throughout most of this early period. And even though these years of development were dominated by Standard Oil (New Jersey), the "Oil Trust" was unable to prevent the entry and growth of competitors (Shell, Gulf, Texaco, Sun, etc.) nor prevent a substantial decline in its own considerable market share. In short, the early years in petroleum were both unregulated *and* competitive with no public crisis in either the price, supply, or distribution of products.[1]

However, the long-term decrease in energy prices which occurred during this period would not have happened if there had been government price controls instead of a free market in which energy producers competed on the basis of price. This is the point Mitchell makes in reference to the substitution for whale oil of kerosene made from petroleum when he says,

> I suppose anyone brought up on the current energy discussion would find it hard to believe that this all happened without an energy policy, and yet it did. Indeed, I think it might well not have happened had there been an energy policy and an energy czar. Because for this resource revolution to occur required the stimulus of a substantial rise in whale oil prices. And even though in the end prices were lower, would a czar have permitted this kind of automatic market adjustment to occur? You can speculate on the possibility of controls on old whale prices and young whale prices, and you can contemplate threats to break up the whaling companies, and you can even envision presidential moralizing about the virtues of darkness. But what's clear is that . . . the kinds of policies that we talk about with regard to the energy industry today would all have frustrated if not sabotaged the transformation that did occur, and that this transformation was in the consumer's interest and in the nation's interest.[2]

The kinds of policies that would have "frustrated if not sabotaged" the interests of consumers and the nation are government price controls which result in prices being compulsorily set above or below free-market prices. Such controls have actually been part of the energy scene for the last half-century. Initially, government controls were used to set prices above free-market levels. In more recent times, controls have been used to set prices below free market levels. In both cases, the results have been disastrous, and the latter approach is a major cause of the current "energy crisis."

The story begins during World War I, when, as Armentano relates, the early laissez faire era in oil abruptly ended. Due to war needs, oil industry executives agreed to cooperate with the government in "emergency" wartime planning, which involved fixing prices, determining outputs and allocating oil supplies. In effect, Armentano points out, the wartime experiment in "planning" created what had previously been unobtainable in the petroleum industry: a governmentally sanctioned cartel.

When the war ended, a "strong sentiment among oil industry leaders existed for continuing the National War Service Committee's spirit of cooperation and 'supervised competition' with respect to the petroleum industry." A new trade association, the American Petroleum Institute, was formed "to afford a means of cooperation with the government in all matters of national concern."[3]

This was a time when it was thought that an oil shortage was imminent. William H. Peterson tells how in 1919 Dr. Van H. Manning, director of the U.S. Bureau of Mines, declared that "oil from the United States will continue to occupy a less and less dominant position, because within the next two to five years the oil fields of this country will reach their maximum production and from that time on we will face an ever-increasing decline." Based on this and other analyses, conservation of oil became the order of the day. The Federal Oil Conservation Board, established in 1924, recommended compulsory withholding of oil resources and stated that the oil depletion rate could be lessened only by tapping larger quantities of foreign petroleum. Directly and indirectly encouraged by government, American oil investments abroad increased from $400 million in 1919 to $1.4 billion in 1939 and spread from Mexico to Venezuela, Iraq, the Dutch East Indies and elsewhere.[4]

But oil prices failed to reflect the "running out" views of the twenties, going down instead of up, as they would have if oil had been getting scarce. The reason oil prices went down was increased production from important oil strikes in West Texas, Oklahoma and California in 1926 and 1927. Then, in 1930 at the beginning of the Great Depression, the giant East Texas oil field was discovered and its yield began flooding the market, driving oil prices still lower. All of a sudden, importing foreign oil became a "no-no." The Independent Petroleum Association of America, established in 1929, appealed for import restrictions. Governors and legislators of major oil-producing states called for an oil tariff. The Revenue Act of 1932 passed with a tax on petroleum imports. And the NRA Petroleum Code of 1933, later declared unconstitutional by the Supreme Court, empowered the president to restrict imports via quotas.[5]

But, although the hysteria about the need for more oil imports was quickly snuffed out, the conservation ethic lived on, even though the need for conservation was for the time at least seemingly obviated by new finds. The Oil States Advisory Committee, formed in 1931, recommended passage of state and national conservation laws, and practically all oil-producing states enacted local conservation statutes if they had not already done so.[6]

The case for conservation presented in many of these statutes was based on a technical problem. It seems that the many different owners with common rights in the large, newly discovered pools of oil all began pumping at a furious pace, fearing that if they didn't somebody else would. This rapid removal threatened to reduce the total amount of oil that might be produced over the life of the fields because it prematurely lowered the well pressures that forced the oil up to the surface. Of course, it also resulted in a glut of oil and a precipitous fall in prices. Since it was in the owners' long-term economic self-interest to maximize total production over the life of their fields, one might suspect that left to their own devices they might eventually have gotten together and worked out some sort of cooperative arrangement. One possibility was field unitization, under which the owners themselves would control output based on the maximum efficient rate of production.

However, state regulatory agencies stepped in instead, imposing mandatory prorating systems among field owners and establishing a system for controlling total state production levels. Ostensibly, the purpose was to conserve oil by insuring maximum production over total field life. But the controls were also used to restrict production and set prices above levels that would otherwise have existed in the free market.[7] In Texas, the controls were enforced by National Guard troops during the early thirties, and East Texas crude oil prices rose from ten cents a barrel in August of 1931 to eighty-five cents a barrel in June 1932.[8] In 1935 Congress proceeded to buttress what was in effect a state-operated oil cartel by passing the Connally Hot Oil Act, which declared as contraband any crude oil produced in excess of a state prorating system and shipped in interstate commerce. The same government which in 1911 broke up the Standard Oil Trust operated by private owners passed a law in 1935 to protect an oil cartel operated by government!

Oil prices established by the government-operated cartel were substantially above free-market levels. But, as Mitchell points out, production costs were also enormously higher because numerous small, high-cost producers with political power were favored with generous production allocations at the expense of larger, lower-cost producers on whom excess costs were imposed. (Stripper wells producing ten barrels of oil or less a day, for example, were exempted from controls, while larger producers were restricted to a fraction of their capacity, resulting in higher production costs per barrel.) "It is certain that numerous inefficient producers were enriched by the cartel," Mitchell states. "The profits of medium-sized and larger producers, however, appear to be substantially below the normal rate of return earned by U.S. industry. . . . For these firms the added costs seem to have more than offset the higher prices." Mitchell goes on to stress that "this cartel was operated by the government and could *only* have been operated by the government. The structure of the crude oil producing industry was and is sufficiently competitive to make a private cartel impossible. Indeed, the major companies would never have designed a system so overwhelmingly hostile to their interests. The production control systems developed appear to be the work of small producers employing their relatively greater influence on state legislatures. . . ."[9]

By 1955 crude oil prices had more than doubled from World War II levels. But imports also increased by four times, due in part to changes in import restrictions for military reasons during the war and government encouragement of major American oil companies to develop foreign oil fields for national security purposes following it. But the chief reason for the increase in foreign imports was economic. Domestic oil had been becoming scarcer and more expensive to produce due to rising discovery and drilling costs, while foreign oil, especially from the Middle East, had been becoming cheaper and more plentiful. No new, major domestic oil fields had been discovered in the United States since the thirties, while many very large fields had been found in Canada, Venezuela, the Middle East and Indonesia.[10] If free access had been allowed to foreign imports at this time, oil prices would have quickly dropped to free market levels and higher-cost domestic oil would have been conserved for future use.

But this is not what happened. Domestic oil producers invoked a "national security" argument. However, they were also concerned that substantial postwar price increases would be washed out by the foreign onslaught and domestic production would have to be cut back to make room for the indefinite expansion of overseas imports. So in 1959 the federal government put into effect the Mandatory Oil Import Program, which restricted imports to a quota fixed as a percentage of domestic production.[11]

The Mandatory Oil Import Program made it possible to maintain domestic oil prices above world free-market prices, providing financial support to the domestic oil industry at the expense of consumers. J. Clayborn La Force, chairman of the department of economics at UCLA, estimates that the import controls cost consumers roughly $5 billion a year in 1969 prices. Domestic oil producers made about $2.7 billion of this in extra revenues. The remaining $2.3 billion were spent by consumers on conservation measures such as insulation, which would have been unnecessary and uneconomic if free-market prices had prevailed.[12] The import program was successful in maintaining artificially controlled high prices for domestically produced oil. As Mitchell points out, in spite of "considerable excess capacity in the domestic producing industry, U.S. crude prices remained virtually constant through the 1960s, typically running $1.25 per barrel (or 60 percent) over the landed foreign cost."[13]

But the oil import program also marked the beginning of the end for the state-operated cartel in oil. For, in the same proclamation establishing the program, President Dwight D. Eisenhower, with one eye undoubtedly on a budding consumer movement, extracted a quid pro quo in the form of a not-so-veiled threat of price controls if domestic oil prices increased, stating that "in the event prices of crude oil or its products or derivatives should be increased after the effective date of this proclamation, [government surveillance of the industry] shall include a determination as to whether such increase or increases are necessary to accomplish the national security objectives of the Act . . . and of this proclamation."[14]

Prices remained constant throughout the sixties, but since this was a time of accelerating inflation, the real price of domestic oil dropped by 15 to 20 percent. This resulted in a crunch on the industry because its high and growing discovery and drilling costs actually required prices to increase if it were to maintain its reserves. Without price increases, reserves and surplus capacity were used up throughout the sixties, interrupted only by the find on the North Slope of Alaska in 1969. At the same time, domestically produced natural gas, its price kept very low under government controls, was also maintaining a competitive lid on domestic oil prices.

The resulting shrinkage in the domestic oil industry is described by C. John Miller, former president of the Independent Petroleum Association of America, who relates that "in 1956, we were drilling about 57,000 wells per year in this country. We had 2,650 active operating rigs. We had 20,000 people like myself, independent oil and gas operators, and we had 2 million barrels a day of surplus producing capacity. By 1971, the number of operating rigs had dropped from 2,650 to 950. Half of the independents had gone out of business. Well drilling had dropped from 57,000 wells to 27,000 wells a year. . . ." Why did this decline occur? "In 1958, my brother and I drilled a well near Ludington, Michigan," says Miller.

"We sold the oil from that well for $2.95 per barrel. Fifteen years later . . . in 1973, we drilled another well. This second well was drilled 2½ times as deep as the first, and it was seven times more expensive to drill. We sold the new oil for $2.95 per barrel. If you are alive and breathing and well, you know it is absurd that anyone in any business went for that period of time without an increase in the price they were receiving for their product. Our labor costs went up as much as in any other business. Our costs of raw materials, land, exploration, drilling and every other part of the business went up. Practically all consumer prices, from sandwiches to college educations, went up dramatically in those years—but not the price of oil."[15] So forty years of government attempts to control domestic oil prices at levels higher than the free market finally ran out of gas, leaving in their wake a multi-billion-dollar oil bill for consumers, depleted oil reserves and a debilitated oil industry.

The story of government controls on natural gas is different from that of oil, but the results are equally disastrous. Government price controls were imposed on the interstate sales of natural gas in 1954, following a Supreme Court decision granting price-setting authority to the Federal Power Commission. The controls in this case were used in an attempt to benefit consumers with low energy prices. The price of natural gas was rising at the time controls were imposed, but natural gas was still underpriced and a tremendous energy bargain in comparison to other higher-priced fuels such as oil and coal. Initially set above free-market levels, the price controls gradually reduced natural gas prices below market as inflation grew in the sixties. This put clamps on natural gas production as a result of rising, inflationary costs which could not be recovered at the low, controlled prices. The number of natural gas wells drilled consequently plunged from 57,111 in 1956 to 28,008 in 1970. Natural gas reserves fell from a twenty-one-year supply in 1956 to less than fourteen years in 1970.[16]

At the same time, consumer demand for natural gas was increasing because its price was artificially controlled below free market levels. The result was costly shortages beginning in 1970 when production fell short of demand. The price controls provided economic benefits in the form of prices below free market levels to residential and industrial consumers who were already using or were lucky enough to get hooked up to use natural gas. But the controls also economically penalized other consumers who would have been more than willing to pay free market prices for natural gas but because of shortages resulting from price controls had to purchase even higher-priced energy such as fuel oil or electricity to heat homes and meet other needs. These non-gas-using consumers were approximately equal in number when weighted by consumption to favored gas-using consumers in 1968. But their number was growing faster. Consequently, it could be anticipated, as Mitchell points out, that more households weighted by consumption would eventually be made worse off than better off by natural gas price controls. In addition, natural gas shortages resulting from controls began causing costly plant layoffs such as one in Cleveland in 1970 which idled 30,000 workers for ten days. Furthermore, the number of plants that were not built and the jobs that were not created in gas-short areas due to price controls will never be known.[17]

In contrast to oil and natural gas prices, coal and uranium prices are not directly controlled by government.

But coal and uranium are primarily used by utility companies to produce electricity. Since these utilities are government-sanctioned monopolies, the price of electricity (as well as natural gas supplied by gas utilities) is government controlled. Control over prices is exercised by federal, state and local regulatory commissions, which establish rates designed to enable electric utilities to recover costs and achieve a predetermined return on investment.

Historically, electric utilities were able to keep electricity prices low through increases in electrical generating efficiency, which made it possible to construct new generating plants at less cost than old ones. As a result, since 1950 retail electricity rates actually went down in some regions of the country, while the overall increase in electricity prices nationwide was only 11 percent in comparison to consumer prices which increased by 50 percent.

But, in the mid-sixties, the trend toward greater efficiency began coming to a halt as maximum technical efficiency with existing steam-powered generating equipment was approached. At the same time, costs continued to rise, requiring that the utilities request larger and more frequent rate hikes from governmental regulatory commissions. But such commissions were historically not used to granting rate increases as large as or as often as required. In addition, the consumer movement was playing a growing role in rate hearings. The upshot was that the electric utilities were increasingly denied sufficient rate increases necessary to insure their capacity to build required future generating capacity.

This resistance to rate hikes provided short-run economic benefits to consumers. But these were increasingly obtained at the cost of possible future electrical shortages, brownouts and blackouts resulting from lack of sufficient generating capacity.

By 1970 the prices of hard energy—domestic oil, natural gas and electricity produced from coal and uranium—were all being controlled by government. The price of high-cost domestic oil, previously pumped up by government to a level above the free market price, had been effectively controlled since 1959, while discovery and drilling costs continued to rise. The price of natural gas, underpriced in the first place, had been controlled below free market levels since the mid-sixties. The price of electricity had also been prevented from rising at a sufficient rate to insure adequate future electrical generating capacity since the mid-sixties. As a result, the prices of all three forms of energy show a pattern of rising during the fifties and flattening out during the sixties:[18]

Year	Crude Oil (cents per million BTUs)	Natural Gas (cents per million BTUs)	Electricity (dollars per 500 kW residential)
1950	43.3	6.0	10.11
1955	47.8	9.8	10.30
1960	49.7	13.6	10.62
1965	49.3	15.1	10.41
1970	54.8	16.6	10.51

But these were also years of inflation, so it is even more important to look at what happened to real, or deflated, prices as a result of government controls. During the fifties, crude oil declined in real price even though its price was supported above free-market levels by the government-operated state cartel. Furthermore, even with import quotas, its real price declined faster during the sixties due to the threat of government price controls and increasing competition from natural gas.

Why natural gas was such a potent competitive threat to oil is shown in the table above. In 1960, for example, the price of natural gas for one million BTUs was 13.6 cents, while the price of crude oil for the same amount of heat energy was 49.7 cents. Being below market in the first place, the real price of natural gas increased substantially during the fifties, but steadied and then slightly decreased during the sixties as a result of controls.

Meanwhile, the real price of electricity decreased throughout the fifties and sixties. These trends in real prices (1972 dollars) are shown below.[19]

Year	Crude Oil (cents per million BTUs)	Natural Gas (cents per million BTUs)	Electricity (dollars per 500 kW residential)
1950	80.7	11.2	18.84
1955	78.4	16.1	16.89
1960	72.4	19.8	15.47
1965	66.3	20.3	14.01
1970	61.5	18.2	11.50

Government price controls thus reduced the real prices of crude oil and electricity and restricted the real price of natural gas throughout the sixties to 1970. All things being equal, such an artificial, government-controlled reduction in the real price of energy or any other commodity will result in less production. But all other things weren't equal. While these energy prices were going down, the costs of producing the energy were going up spectacularly by 1970, as inflation got into full swing, further reducing incentives to expand production. Meanwhile, as if this weren't enough, the Conserver Cult was hitting energy production with a double whammy. First, consumerists were throwing their substantial political weight behind not only maintaining but expanding controls over energy prices. At the same time, the cult's environmental wing was using environmental and safety requirements to rapidly increase energy costs. For example, in the case of oil, during the short period from 1967 to 1971 emission control equipment on automobiles sharply increased gasoline consumption; the Alaska pipeline was delayed for five years due to environmentalist legal challenges; the oil spill off Santa Barbara in 1969 prompted a four-year moratorium on California offshore drilling; oil refinery construction was repeatedly delayed or abandoned altogether because of environmentalist pressures; and, most important, the Clean Air Act and various state laws restricting sulfur emissions prompted a massive industry shift from high-sulfur oil to more expensive low-sulfur oil, especially by utilities in the Northeast.[20] At the same time, other environmental and safety costs were piled up on coal and nuclear power production and costs of developing nuclear powerplants began skyrocketing because of time-consuming and expensive licensing delays.

It is therefore no wonder that production of hard energy peaked in 1970, caught in a Conserver Cult squeeze of price controls and rapidly rising costs. It is also of little wonder that the United States, its growth in energy production cranked down to zero, became a sitting duck for the Organization of Petroleum Exporting Countries. OPEC was spawned in 1960 primarily as a result of the Mandatory Oil Import Program legislated in 1959 to restrict U.S. imports of foreign oil.[21] Limited in access to the huge United States market for oil by the program, foreign governments found themselves with an increasing glut of oil and falling world oil prices. They increasingly began taking power into their own hands by banding together under the OPEC banner and eventually taking over control of their own fields. As Armentano explains, although American oil companies held important "concessions" abroad, host governments increasingly decided to withdraw a portion and eventually all of these so-called concessionary privileges. They demanded and received an increase in their "royalties" and then, in many important producing areas (Saudi Arabia in particular) assumed strong national control over crude oil production.[22]

With growth in world demand for oil, the OPEC nations were able to raise prices by forming a cartel to restrict production, (the same thing our state and federal governments did with domestic oil in the thirties). This development may have been inevitable because of the increasing cohesiveness of OPEC members. But Mitchell points out that some observers maintain that "there was nothing inevitable about the turnaround in prices. Rather, it was the result of incompetent U.S. State Department dealings with a highly fragile cartel. Had the U.S. not passively accepted Libyan production cutbacks in 1970, nor given what was tantamount to outright endorsement of higher prices in Teheran in 1971, the OPEC nations would have returned to the competitive pricing patterns of the past two decades and prices would have continued to fall."[23]

But that's not what happened. Under the leadership of Henry Kissinger, who as secretary of state used airplane fuel as if it were going out of style but after leaving office became a fervent promoter of energy conservation, the

United States government effectively acquiesced in the takeover of world oil prices by the OPEC cartel. At this point, the federal government could have simultaneously made the decision to start increasing energy production by eliminating controls. Instead, the Nixon administration, which since 1969 had been publicly jawboning the oil industry to keep oil prices down, put oil as well as all other goods and services under price controls. Then, in 1973, controls were removed from everything but oil and other energy sources. The result was that, with domestic oil production not only stymied but decreasing and demand for oil continuing to increase, OPEC was not only able to up its prices and make them stick but effectively to stage an oil embargo during the Yom Kippur War. Nor did even the costly shortages and expensive chaos caused by the embargo result in moves to decontrol oil and other energy prices in order to increase energy production. In fact, just the opposite happened. Price controls were retained on domestic energy production, which was decreasing because of the controls, in order to protect consumers from the rising prices of OPEC oil imports, which were increasing because controls were causing domestic oil production to decrease. This, of course, was insane. It created an economic opportunity for OPEC to increase its prices even faster.

According to William E. Simon, secretary of the treasury at the time, it was due to the fact that liberals in Congress are "intellectual basket cases in the realm of basic economics. It takes an immense resistance to logic and fact not to know that one cannot simultaneously control prices, inflate costs, ban production, increase taxes, grant counterproductive subsidies—and expect healthy, vigorous production to result."[24]

Simon is no kinder to conservatives, pointing out that

in 1975 President Gerald R. Ford recommended the immediate lifting of all price controls from oil and natural gas since it is "impossible to have increased exploration, production and innovative technology when the government holds the price artificially below the market price." But the Democrats resisted and Ford finally signed into law a forty-month extension of price controls because White House political advisers feared he would otherwise lose the New Hampshire presidential primary. "So principled were these advisers," comments Simon, "that I am convinced, if Texas had been the nation's first primary, these men would have been urging a veto. As it was, the President finally ended up agreeing to *more complex* price controls on *more* oil than had ever existed before. His subsequent pleas that Congress deregulate natural gas were simply ignored, and the nation was saddled with a disastrous energy law."[25]

When Jimmy Carter was campaigning for the presidency, he sent a letter to the governors of gas- and oil-rich Texas, Louisiana and Oklahoma promising to "work with Congress, as the Ford administration has been unable to do, to deregulate new natural gas. The decontrol of producers' prices for new natural gas would provide an incentive for new exploration and would help our nation's oil and gas operators attract needed capital."[26] But once in office, Carter fiercely opposed deregulation of natural gas prices. In 1978, a compromise bill called the National Gas Policy Act was finally passed by Congress and signed by the president. This legislation ostensibly promises to gradually reduce and by 1985 eliminate price controls on newly discovered natural gas. But the act simultaneously extended price controls to cover not only interstate but intrastate gas, which previously had been uncontrolled. In addition, it set up so many new and different classifications and methods of pricing for new and old gas that its result is expected to be an explosion in regulations, litigation and red tape rather than an increase in natural gas production. In 1979 President Carter acted to completely decontrol crude oil prices by 1981. But he coupled this with a call for a "windfall profits" tax that would effectively pull the rug out from under any increase in oil production due to decontrol because it would take revenues that oil companies could otherwise use for increased discovery and drilling operations and give them to the government instead.

However, even if price controls continue to keep a lid on domestic energy production and result in energy shortages, they at least hold down costs to consumers, right? Wrong. Controls on natural gas, for example, do reduce prices to a minority of consumers. But this economic benefit is achieved at the cost of shortages such as the one in the winter of 1977 when thousands of factories had to be shut down, millions of people were thrown out of work and even homes and schools went without heat. And these costs will grow in the future if price controls continue to depress the production of natural gas. For the nine-year period from 1977–1985, these additional costs may total as much as $180 billion— primarily in higher consumer prices for more expensive alternate fuels and gross national product losses due to shortages—according to Robert S. Pindyck, associate professor of economics at MIT's Sloan School of Management.[27] By contrast, truly effective phased deregulation of natural gas prices would result in a total cost due to higher prices of $54 billion over the same period. In other words, the American people as a whole will pay the difference between $180 and $54 billion, $126 billion, during this nine-year period for the "protection" against rising prices supposedly provided by natural gas price controls. Meanwhile, in the case of oil, price controls are resulting in a drain on the economy in the neighborhood of $50 billion a year due to the need to purchase OPEC oil at cartel prices, and this bill can only go higher in the future.

But, if price controls result in increased rather than decreased costs to consumers and the country just as production controls and import quotas imposed by government did in the past, why have them? It does seem passing strange that we follow an energy policy in which, as John J. McKetta pointed out in 1979, we "pay our own U.S. producers only $1.90/million BTUs for natural gas (but only for newly discovered gas—most of the old gas sells for as low as 50¢) but offer as much as $4.35 for imported liquefied natural gas (LNG). We pay our own U.S. producers $12.50 for new oil ($5.60 for old oil) but pay more than $15.00 for imported oil."[28]

Initially, the Carter administration attempted to claim that no additional domestic oil and natural gas were available and that, therefore, production would not rise regardless of how much prices increased. But the government's own research gave the lie to this claim. President Carter then accused oil companies of "profiteering," claiming that oil decontrol would result in "windfall profits" and that the 1976 oil profits of the 30 top oil companies already totaled more than $75 billion. But, as McKetta pointed out, it is "embarrassing to have our president mislead the nation intentionally in order to try to sell an inferior energy plan. The true figure for the net profit of the top 35 [emphasis added] oil companies is $11.4 billion, from which these companies paid $4.2 billion dividends to their stockholders, leaving only $7.2

billion to be reinvested in further exploration, drilling, etc. The U.S. government received $53.8 billion or 72 percent of the income (President Carter neglected to mention this)."[29] And the oil companies' 1976 net profit of $11.4 billion, it turns out, is less than the Carter administration's Department of Energy budget of $12.6 billion. This amazed Congressman Jim Collins, who exclaimed that "the Carter energy plan costs more money for bureaucracy and planning than the oil and gas industry spent in getting their production job done."[30]

Furthermore, although prices of domestic oil and gas have risen dramatically as controls have been periodically raised since 1970, so also have costs as inflation has continued to build up steam. Consequently, Citibank's petroleum department could report in 1978 that, as of 1977, the petroleum industry's rate of return on net worth was "not out of line. In fact, over the 21 years plotted, petroleum's return actually has averaged fractionally lower than that of other manufacturing. And last year, compared to petroleum's rate of return of 14% nonessential industries such as soft drinks, tobacco and cosmetics had returns in the 19% to 23% range. In view of this, it is difficult to fathom the reasons why government restraints—economic and regulatory—remain in place to hamper the petroleum industry. These prevent this vital industry from achieving a level of income, and thus a rate of return, consonant with its importance to the efficient working of our economy."[31]

Another charge made is that there is no free market for oil and natural gas because production is dominated by a domestic cartel of a few large companies. But the previous domestic cartel, made possible only by the intervention of government, became inoperative when world oil prices passed up domestic prices and import quotas were abandoned in the early 1970s. Moreover, as former Congressman and Secretary of Defense Melvin R. Laird observes, natural gas supply is "one of the least concentrated natural resources industries in the country. The largest ten firms account for less than 40 percent of the market, and small independent wildcatters abound. Interstate distribution and retail sales are already the province of fully regulated public utilities. With regard to oil, the major companies are indeed enormous and include some of the largest corporations in the world. However, the degree of concentration of the market is much less than for other major sections of the U.S. economy, such as automobiles, steel, copper, and aluminum. If the fear is concerted action and conspiracy, the remedy is enforceable disclosure and antitrust laws to provide assurance and remedies."[32] More to the point, it is not a private cartel but the monopoly power of the federal government to fix oil prices below market levels and thus restrict domestic energy production which is making it possible for the OPEC foreign oil cartel to successfully operate in the United States today.

Why, then, control prices below market levels? Or, as Milton Friedman asks, why is it that we have "adopted these counterproductive controls on the prices of gas and oil, that we have enacted this enormously expensive boondoggle of a Department of Energy?" According to Friedman, the "answer is straightforward. The enactment of such intervention has been and remains politically profitable. . . . [Politicians] are trying to acquire votes. They are trying to acquire position and influence, and they will do what's politically profitable to do. They have made political hay by professing to protect consumers. . . ."[33]

But, in addition, politicians have made "political hay"

by doing what they perceive to be the bidding of the Conserver Cult. For price controls ultimately result in higher energy prices to consumers, which reduces not only energy demand and consumption but the environmental effects of energy use. At the same time, price controls result in reduced domestic energy production and increased energy imports, effectively "exporting" pollution resulting from production to other countries. Most important, price controls result in energy crises which can be used for political propaganda purposes to gain greater power over domestic energy production. UCLA economics professor Armen Alchian makes this point, stating that "crises are created by politically imposed price controls. When the controls become effective and attenuate the operation of the price-directed market system, complaints are made that the private sector is inherently defective and should be subjected to still greater political control—as witness the politically created energy problem."[34]

The most striking example of this political power play to gain greater control over energy production is President Carter's proposal to decontrol oil prices while taxing away so-called windfall profits of $88 billion resulting from decontrol to fund a federal "Energy Security Corporation." The stimulative effect of this proposal on oil production would be nil because the profits that might otherwise be invested in achieving increased oil production would be taxed away. And it is not like the money isn't needed, according to a 1979 Chase Manhattan Bank study which showed that over the latest five available years some thirty leading oil companies had made capital and exploration outlays of $126 billion, which exceeded net income by $59 billion, or 88 percent.[35]

What the Carter proposal actually would do is increase

the price of oil to consumers and block future reductions in oil prices by eliminating the increased production which would otherwise result from decontrol. Who thinks this is a good idea? The Sierra Club, that's who. Environmentalists for the most part have opposed price decontrol. But, prior to Carter's proposal, several environmentalist groups, including the Sierra Club, urged him to decontrol the price of crude oil and put a tax on oil industry profits that would be rebated to consumers, particularly those on low or fixed incomes (no one explained how this would help these consumers if there were no energy to buy). Jonathan Gibson, an official of the Sierra Club who asked other prominent environmental organizations to cosign a letter to the president, explained why, stating that this could indirectly aid the environment by forcing consumers to use fuel more sparingly. If that happened, he said, auto pollution, strip-mining damage, oil spills and other environmental problems associated with heavy energy consumption should be eased.[36]

"Public interest groups" such as the Citizens Tax Reform Research Group—a Ralph Nader organization—and the Consumer Federation of America are also fighting to either stop decontrol or pass a windfall profits tax with decontrol.[37] A new Nader-led coalition named Consumers Opposed to Inflation (COIN) consisting of seventy consumer, labor, church and senior citizen groups, has targeted energy as the key inflation culprit, calling for creation of a government energy corporation to compete directly with the major oil companies by exploring for and producing oil on federal lands.[38]

Following these environmentalist/consumerist recommendations, Carter's scheme would tax away $88 billion in windfall profits and give these funds to the proposed

Energy Security Corporation ("windfall taxes" are okay, but not windfall profits). What would the Energy Security Corporation do with the money? When Carter first discussed this question, he said the bulk of the money would go for developing new, economically high-risk sources of energy, stating that "we can make it economical to harness the energy of the sun, the wind, the tides and the geothermal power of the earth."[39] Later, he changed the story to read that the money would be used to boost development of an industry capable of producing the equivalent of 2.5 million barrels of oil a day by 1990 in synthetic fuels and natural gas from "unconventional sources."[40]

Aside from the fact that it would do nothing to increase energy production for ten years—which perhaps is its real purpose—the proposal is first of all an extremely expensive way to go. The equivalent of an additional 2.5 million barrels of oil a day can be gotten a lot more cheaply and a lot earlier than 1990 simply by leaving the money in the hands of oil companies to invest in extracting real crude oil and natural gas from the ground, eliminating the costly added operation of synthesizing these fuels from coal or other energy sources. The proposal is also unnecessary since, as MIT economics professors Paul L. Jaskow and Robert S. Pindyck, point out, there are "really only two major impediments in the private commercialization of new energy technologies—the uncertainties associated with possible future price regulation or taxation, and uncertainties about environmental restrictions. . . . What is needed is the removal of controls on the current *and future* prices of energy supplies. This, together with a revision of those environmental regulations that are unnecessary and unreasonable, and the clarification of environmental standards and regulations that would apply to the future, would permit private firms to develop economical new energy technologies efficiently and without any need for government subsidies."[91]

Carter's proposal, in fact, fails to achieve any goal except one: increased government and Conserver Cult power over energy production. It would be naive to think that the precedent of taxing away so-called windfall profits from oil companies might not be expanded at a later date to include the profits of all energy companies. The fact is that, in one sense, all the profits of all energy companies are windfall profits. This is what Douglas Fraser, president of the United Auto Workers union—which is now in the business of protesting energy prices, shortages and "excessive" oil company profits and is pondering participation in nationwide consumer boycotts—means when he says, "Exxon and Gulf and Mobil didn't put that oil in the ground."[42] But neither for that matter did natural gas companies put that natural gas in the ground or coal companies put that coal in the ground or uranium companies put that uranium in the ground.

The proposal for an Energy Security Corporation is only the first step in making energy production in the future dependent on what University of Michigan business administration professor and former Council of Economic Advisers Chairman Paul W. McCracken calls a "nationalized-industry strategy for bringing new sources of energy to the market in the longer run. This is the meaning of the proposal for an Energy Security Corporation to be financed apparently outside the budget. . . . We are asked, in short, to depend for our future energy supplies on a Post Office corporation approach."[43] To put it more plainly, and sticking with the Post Office example, there would be less and less hard energy at higher and higher prices, a Conserver Cult dream come true.

But the real price of Carter's Conserver Cult program to tax away windfall profits and create an Energy Security Corporation is even higher. For what, in fact, are windfall profits? Are the profits homeowners have made over the past ten to fifteen years in rapid appreciation of their house values windfall profits? They most certainly are, according to Carter's definition of such profits as profits "unrelated to [the beneficiaries'] activities or economic contributions." Could they then be taxed away? They most certainly could be. As La Force points out, the "astonishing principle enunciated here is that when the value of an asset rises because of circumstances unrelated to the immediate activities of its owner, that increased value belongs, not to the asset's owner, but to the government! . . . We should be guilty of egregious naivete to believe that the value of anyone's private property is safe from confiscation—now that government openly and deliberately has started down this path. While those who would dramatically restructure our society have long pushed this principle, Jimmy Carter is the first President of the United States to embrace it openly and to weave it as a central theme into a critical domestic program."[44]

In other words, the price of the Conserver Cult utopia is not only the throttling of future energy production but the end of our free society.

14

ENERGY CONSERVATION

Energy use has grown as the price of energy has declined over time. This has made it possible to increasingly conserve the most expensive and important energy resource of all—human energy. It has also made it increasingly possible to conserve the only truly unrenewable resource—time. The result has been continuing growth in gross national products (GNP) and per capita income with consequently better and better living standards. As Aden and Marjorie Meinel put it, cheap energy is "truly the basis upon which modern society is built and which frees us from the severity of a dispassionate environment. Agriculture and fire set the basis for cities, but since only slaves and animals were available for motive power and industry the lot of most humans was rather dismal. The harnessing of energy and inventions, and laws to control their exploitation, opened the way to freedom and dignity."[1]

In the United States, the role of increasing supplies and new types of energy in not only spurring economic growth but removing constraints which otherwise would have severely impeded the rate and diversity of economic development is described by Sam H. Schurr, codirector of the Center for Energy Policy Research of Resources for the Future:

As late as 1870, about three-quarters of all the energy used in the United States was still coming from fuel wood, but the transition to coal was under way and coal soon became the dominant source. What was of primary significance in this transition was not that coal could substitute for wood in existing energy uses but rather that this was a change from a fuel resource severely limited in supply to another available in apparently endless amounts. The use of coal thus opened the way for the large-scale, unimpeded growth of iron and steel production. Ample supplies of iron and steel, in turn, made it possible to build and operate a railroad network that blanketed the country and to produce the machines required for the expansion of manufacturing. Once the fuel constraint was broken, one development led to another in a dynamic sequence that laid the foundation for modern industrial society.

In the twentieth century, as the composition of energy supply moved toward liquid fuels and electricity, other major constraints to economic growth were overcome. Electricity and the electric motor removed the limitations imposed on factory processes by the earlier mechanical energy systems, which used shafts and belting to transmit power from the in-house prime mover. Through the reorganization of factory production, which they made possible, electricity and the electric motor paved the way for large-scale productivity increases in manufacturing. Liquid fuels, the tractor, and energy-based fertilizers led to enormous increases in crop yields by removing the limits that had been imposed on agriculture by the availability of natural fertilizers and by animal draft power. And as agricultural productivity rose, farm workers became available for other sectors of production.

Geographic constraints also were removed through developments in energy supply. During the nineteenth century, railroad transportation and the mobility of coal removed the strict limits formerly imposed on industrial locations by waterways required for transportation and water wheels needed for mechanical power. In the twentieth century, the truck and the automobile broadened the availability of transportation routes, and the coming of liquid and gaseous fuels further increased the mobility of fuels, thereby removing the constraints on industrial locations previously imposed by railroads and coal. More recently, air conditioning and air transportation have removed other limitations to economic growth in many regions of the United States and the world.

The removal of all these constraints has resulted in national, regional, and local development and growth; increased productivity in industry, agriculture and transportation; greater production of goods and services for human consumption; and marked improvements in personal living comforts and amenities. It is important to observe that in these adaptations, energy was not substituted marginally for other factors of production, such as capital and labor, which could have produced essentially the same final outcomes. Instead, energy supply and associated technologies made practical by developments in energy supply together produced results that could not have been achieved in other ways.[2]

Schurr points out that the growth of energy consumption and GNP can be divided into three distinct periods. First, there was an early period extending from the latter half of the nineteenth century to about the second decade of the twentieth century when energy consumption grew at a rate faster than GNP. Then, there was a middle period from about the end of World War I to mid-century during which energy consumption grew at a slower rate than GNP. Finally, there is the current period from mid-century to the present in which energy consumption and GNP have grown at approximately the same rate. Schurr notes that "despite the close relationship between GNP and energy consumption since World War II, the long historical record does not support the view that these factors have grown essentially at the same rate. Not only did they not grow at the same rate, but their comparative rates show divergence in different directions, depending on the period of U.S. economic history being covered."[3] However, the reason why energy consumption grew faster than GNP in the first period but slower in the second period is that the technical efficiency of energy conversion was increasing at different rates throughout these periods. The Energy Research Group analyzed this energy efficiency over the past century and found that it increased slowly from 9 to 13 percent during the period from 1880 to 1910, then speeded up dramatically to 30 percent by 1950, and finally slowed down again to a current efficiency of about 36 percent:[4]

Technical Efficiency
of U.S. Energy System, 1880–1980

First Period	Second Period	Third Period
1880 – 9%	1920 – 16%	1960 – 32%
1890 – 10	1930 – 21	1970 – 34
1900 – 11	1940 – 27	1980 – 36
1910 – 13	1950 – 30	

Schurr points out that technical efficiency increased rapidly during the second period from 1920 to 1950 because "there were fundamental changes in the composition of U.S. energy output, including a phenomenal rise in the importance of energy in the form of electricity and liquids as opposed to the earlier heavy domination of coal. And there were sharp increases in thermal efficiency in such major areas of energy use as railroad transportation, electric power generation, and space heating.[5]

As the Energy Research Group notes, it is the energy output resulting from the efficiency with which energy is consumed that should bear the closest relationship to GNP. The group developed this ratio of energy output to GNP for the years 1925 to 1970 and found that it "increased during the continuing industrialization of the first third of the century" and "then flattened out to a fairly constant value during the second third," adding that "the ratio of energy output to Gross Private Domestic Product appears even more constant in the 1933–1966 period."[6] This pattern continued through the seventies, when there was a slight decline in energy consumption relative to GNP but a slight increase in energy efficiency, resulting in energy output maintaining a constant relationship to GNP.[7]

Furthermore, there is an even closer constant relationship between per capita energy consumption and per capita disposable income in recent years, according to H. R. Linden, president of the Institute of Gas Technology, who comments that "even more striking is the amazingly precise correlation between total primary energy consumption per capita and disposable income per capita from 1929 to 1973. . . . Regardless of energy prices, employment level, wartime or peacetime, boom, recession or depression, when disposable personal income went up, so did energy consumption in a quantitatively related way. And, when disposable personal income went down, down went energy consumption in the same quantitatively related way."[8]

An extensive historical record stretching over a century thus exists to indicate that there is a close relationship between increases in energy consumption and growth in GNP and per capita disposable income. Initially, due to the heavy energy requirements of industrialization, energy consumption grew faster than GNP from 1880 to 1910 or 1920. Then energy consumption grew more slowly than GNP until about 1950, as improvements in energy efficiency made it possible to achieve more output per unit of input. Finally, for the past three decades, there has been a relatively constant relationship between energy use and GNP, with consumption decreasing slightly while efficiency increased slightly.

Throughout this period, conservation took place in terms of improving the efficiency of energy conversion from a low of 9 percent in 1880 to a current level of about 36 percent. This energy conservation was achieved not by government mandate but solely as a result of successful attempts by energy producers to reduce energy costs in order to be more price-competitive in the marketplace. At the same time, energy conservation was achieved by users based on the price of energy in relation to other goods and services. All other things being equal, consumers tended to use more energy when energy prices fell and conserve or use less energy when they rose. In other words, the marketplace through the medium of energy prices provided both producers and consumers with automatic incentives to conserve by respectively producing or using energy more efficiently or using less energy—which basically are the only methods of energy conservation available.

However, the Conserver Cult is not happy with this approach to energy conservation because it leaves both energy producers and consumers free to respond in their own way to rises and falls in energy prices. So, in the case of energy producers, the cult turned off growth in energy production by imposing stringent environmental laws, absolute safety standards and economically strangling price controls. Then, when an "energy crisis" occurred due to the shortages which inevitably developed, the cult attacked energy producers for "profiteering" and pushed for total power over energy production. But energy production cannot be stopped without also eventually turning off energy consumption. So the cult used the same, self-made shortages to push for total power over energy consumption, attacking consumers as "wasteful energy pigs."

President Carter, who remarked before his inauguration that he was determined to "reform" U.S. energy habits "even if it costs me another term,"[9] led this attack when he introduced his National Energy Plan in 1977, declaring that "ours is the most wasteful nation on earth. . . . With about the same standard of living, we use twice as much energy per person as do other countries like Germany, Japan, and Sweden."[10] Energy Secretary Schlesinger backed this up, stating that "Americans are energy junkies who must be weaned."[11] All of this was right up the alley of Ralph Nader, who opined that the U.S. has a waste factor of more than 50 percent, "which means if we were efficient we could continue economic growth for the next 35 years without increasing the absolute amount of gas, coal and oil we are consuming now."[12] And, of course, none of this was anything that hadn't already been said even more grandiosely by Amory Lovins when he announced in his famous *Foreign Affairs* article that "technical fixes *alone* in the United States could probably improve energy efficiency by a factor of at least three or four. A recent review of specific practical measures cogently argues that with only those technical fixes that could be implemented by about the turn of the century, we could nearly double the efficiency with which we use energy."[13]

What about the charge that "with about the same

standard of living, we use twice as much energy per person as do other countries like Germany, Japan and Sweden"? First of all, it simply isn't true that the United States and these other countries have "about the same standard of living," as anyone who has traveled through any of these countries can testify. Such statements are usually derived using foreign exchange rates for comparative purposes. But these rates are misleading when used to compare standards of living because they are heavily influenced by prices of exported goods, expectations of future inflation, and day-to-day financial speculation. To provide more realistic comparisons of economic conditions around the world, a concept of purchasing power parity in which the currencies of different countries are compared in terms of their actual purchasing power was developed by a team of economists headed by Irving B. Kravis, a University of Pennsylvania professor who directed the research on behalf of the United Nations. Estimates for 1970 and latest years (mainly 1978) are shown in the table below, which expresses living standards for five countries as percentages of the U.S. 100 percent level:

Country	1970	Latest
United States	100%	100%
France	66	68
West Germany	65	66
Britain	63	57
Japan	49	56
Italy	48	46

The data show that the U.S. has a living standard which in terms of actual purchasing power of goods and services is 50 percent higher than the next two countries, France and Germany. A similar finding emerges from statistics gathered by the Paris-based Organization for Economic Cooperation and Development. These data show that "for every 1,000 persons in America, there are 505 automobiles, 571 television sets and 721 telephones. Comparable numbers abroad include: Britain, 255 cars, 320 TV sets and 394 phones; France, 300 cars, 268 TV sets and 293 phones; West Germany, 308 cars, 306 TV sets and 426 phones; Italy, 283 cars, 213 TV sets and 271 phones."[14]

Further, comparing energy usage between countries is like comparing apples and oranges because each country uses energy in different ways. This is why a workshop sponsored by Resources for the Future and the Electric Power Research Institute to study differences in ratios of energy consumption to gross domestic product (GDP) reached near-unanimous agreement in concluding that

the aggregate energy/GDP [ratio] when compared *among countries* is unsuitable as a guideline for policy formulation, since it is an unsatisfactory measure of either economic efficiency or energy efficiency. With regard to economic efficiency, the energy/GDP ratio is unsatisfactory because energy is only one input—and quantitatively a small one at that—into the total of goods and services which represents GDP. Capital, labor, and other materials inputs are much larger. Societies, industries, and firms combine these factors in different ways. The amount of energy inputs or consumption in a country will be affected by the price of energy relative to the prices of the other inputs which can be substituted for energy and other factors such as differences in technology, resource base, and industrial mix. There is, in short, no universally "right" amount of energy consumption, only that which is appropriate to the configuration of economic and other factors in a particular country. Neither is the overall energy/GDP ratio a good indicator of energy efficiency, since the amount of energy consumed in a country relative to GDP is affected by the industrial composition of the economy. A country whose economic activities are concentrated in energy-intensive industry will consume more energy than a country whose economic activities are concentrated in agriculture. In both cases, energy may be used with exemplary efficiency throughout the economy, but this will not stop the energy/GDP ratio of the industrial economy from being much higher than that of the agricultural economy. On these grounds there was near unanimous agreement that the overall energy/GDP ratio when compared *among countries* was misleading and certainly did not provide good guidance for conservation policy.[15]

Invidious energy comparisons are most often made between the United States and Sweden and Switzerland. Lovins, for example, cites studies which supposedly show that, if America were as efficient as Sweden, we could save at least a third of our energy. But, as Peter J. Brennan points out, if

America were more like Sweden, we wouldn't export as much or continue to feed so much of the world. The Swedes use on the order of 40% less energy per person than Americans do, which apparently makes us energy gluttons in [Lovins'] eyes. Lovins' line of reasoning fails to take note of the simple fact that a large portion of the U.S. economy is geared for *export,* while Sweden's is not. Thus, with 5.3% of the world's population, the U. S. produces 17% of the world's grains and cereals, 23.6% of the world's manufactured goods, and 26.6% of the gross world product. This accounts in large part for the discrepancy between Swedish and American per capita energy use. Indeed, in the area of agriculture alone, which accounts for nearly one-fifth of U.S. energy consumption, we have more land under production than the entire area occupied by Sweden! Also, Sweden is neither a large nor suburban-oriented nation. A big chunk of per capita energy consumption by Americans is for gasoline. That is not the case in Sweden. While there is certainly energy waste in this country that can and should be eliminated, simple-minded and muddle-headed comparisons of non-comparable countries make for dramatic but useless conclusions.[16]

Or, as Petr Beckmann puts it,

Sweden and Switzerland have a standard of life almost as high as the U.S., say the anti-nuclear crusaders, yet their per capita energy consumption is only half as great. From which it follows with iron logic that nuclear or any other additional power is unnecessary in the U.S. The Swedes and Swiss are living on half, aren't they? Sure. And the U.S. could, too. First and foremost, it would have to stop feeding itself and much of the rest of the world; it would have to import most of its food like Sweden and Switzerland. . . . Second, it would have to give up its energy-intensive industries, not only agriculture, but also steel, aluminum, copper, mining, smelting, metal working, petrochemicals, paper, and a host of others; it could always turn to low-energy and services industries like Switzerland—watches and banking. Third, the U.S. would have to give up its automobile-based economy. Most Swedes and Swiss use the automobile for comfort and pleasure, not for essential transportation. This would mean, among other slight adjustments, reducing the area of U.S. cities, but not their population, so as to achieve the much higher population density of Geneva, Zurich or Stockholm. What could be easier? Fourth, it would have to arrange, as Sweden and Switzerland did, for almost exclusively mountainous country with a multitude of short, swift, abundant streams that can be harnessed for highly efficient hydropower. The great majority of Sweden's electric power is hydroelectric. . . . Similarly, Switzerland is Europe's most mountainous country, and its abundant streams power no less than 431 major hydroelectric plants.[17]

In other words, if one country specializes in heavy manufacturing while another country specializes in service industries, the first country is going to use more energy per dollar of GNP than the second although both may be equally efficient in energy use. But suppose one could eliminate these variations in the economic bases of different countries and simply consider how much energy it currently takes each to achieve a given increase in GNP.

269

This would be a dynamic rather than static comparison of ongoing energy efficiencies among countries. This is what Aden and Marjorie Meinel did, reporting that

there exists a widespread impression that Sweden, in particular, has discovered a way of having a high per capita GNP with low energy consumption. [Our data] belie this myth. . . .
It takes almost as much energy to generate a 1,000-dollar increment of new GNP in Sweden as it does in the United States. The dynamic approach . . . removes the differences in the economic base of each country and directly addresses the efficiency of each in generating new income. . . . Sweden, Canada and the United States show approximately equal energy efficiency, while Japan and Germany are distinctly better. The United Kingdom and Mexico are best. It is interesting to note that no country has been able to produce an increase in income without the expenditure of an additional increment of energy. There are a few years where significant income gains were accompanied by small increases or even decreases in energy consumption, but these years are rare and generally indicate that a downturn in the economy has already begun. . . . The ideal expressed by prophets of the so-called "Post Industrial Age" is that nations will learn how to reduce energy consumption without significantly reducing living standards. If the standard of living is closely associated with per capita GNP, then it is clear that someone has yet to discover the secret needed to accomplish this new economic performance. . . . It would be nice to have another epoch of increased growth in technological efficiency, as in the 1917–1950 period. But this is not likely to recur, because most of the room for improvement was depleted at that time. Future gains are apt to be through conservation and waste heat utilization, but at an added cost that may make the net results disappointingly small when viewed in a GNP/energy diagram. . . . One thing is clear: no nation has discovered how to reduce energy consumption and at the same time increase or maintain income on more than an accidental basis. History has shown that a small reduction in energy use can be accomplished when there is a corresponding increase in the efficiency of energy use, as happened during the 1920s. Halts in growth can be serious. If

economic growth is halted by internal forces, any prolonged stagnation can lead to internal political and social problems. If that growth is halted by external forces, then serious international consequences can be set in motion that would almost certainly bode ill for the prospects of peace.[18]

But the Conserver Cult still proclaims that energy efficiency can be radically increased without affecting GNP or our standard of living. For example, when Denis Hayes was with the environmentalist Worldwatch Institute, he confidently declared that "more than one-half the current U.S. energy budget is waste. For the next quarter century the United States could meet all its new energy needs simply by improving the efficiency of existing uses. . . . Energy derived from conservation would be safer, more reliable and less polluting than energy from any other source."[19] And the Council on Environmental Quality has weighed in with a statement hearteningly called "The Good News About Energy," in which it claims that "increases in the productive efficiency of energy possible with today's technology would allow the U.S. economy to operate on 30–40 percent less energy. As a result of this large potential for energy savings, we can fuel the growth of the economy in years ahead in large part by increasing the productivity of the energy we now use rather than by greatly increasing our energy inputs."[20]

But the United States has done quite a bit to conserve energy in the past, as is shown by the increase in energy efficiency from 9 to 36 percent over the last century. It is also the case that the most significant increase in energy efficiency which occurred prior to 1950 took close to three decades to achieve. It is for these reasons that Schurr comments that

the changes over time in the relationship between energy and

271

GNP have not been drastic. Even over the several decades when energy consumption declined persistently relative to GNP, the reduction in the ratio came to no more than about one-third. . . . Yet the one-third decline in the ratio between energy consumption and GNP, though significant, was far less drastic than some of the forecasts now being made for the next 25–35 years. . . . The profound changes in energy supply and use technology that occurred during the single sustained period of decline in energy consumption relative to GNP (1920–1945) place a strong burden of proof on forecasts that project a substantial decoupling of energy consumption and GNP in the future. To be credible, such forecasts should be accompanied by specifications of those changes in energy production and utilization technology that are supposed to produce such a decoupling and by plausible evidence of their technical and economic feasibility.[21]

Or, as Fred Hoyle puts it, it "is not difficult to see how we all might save a little energy, but it is a mistake to think that conservation could go very far towards mitigating our future need for energy. Conservation could moderate the need, but not by any great margin, for if *the margin were great, conservation would have happened already.*"[22]

It should also be noted that during the past century when energy efficiency quadrupled from 9 to 36 percent, energy consumption still increased at historic rates of 3 to 4 percent a year. Consequently, if increases in energy efficiency are included, the increase in the rate of actual energy use was considerably higher than 3 to 4 percent. For example, suppose that energy consumption increases from 100 to 200 units during a given period. That's an increase of 100 percent. But suppose during the same period that energy efficiency increases from 10 to 20 percent. This means that actual energy effectively used increased from 0.10 times 100 units, or 10 units, at the start of the period to 0.20 times 200 units, or 40 units, at the end of the period. So the increase in actual energy use is 400 percent, not 100 percent. This kind of dynamic, multiplying effect of increasing energy efficiency on increasing energy consumption resulting in even faster growth of actual energy use is exactly what has occurred in the past and is occurring today. However, this dynamic relationship between energy efficiency and consumption is ignored by the Conserver Cult, which acts as if it alone produced the concept of efficiency out of thin air, conveniently neglecting to mention the continuing efficiency improvements of the past. Thus, the cult proclaims that increases in efficiency will reduce future energy consumption when in fact consumption has increased along with or even in spite of comparable efficiency improvements in the past.

It is for all of these reasons that Ian Forbes of the Energy Research Group states that "there is much room for efficiency improvement, but it can only come at a steady pace. . . . Nor should it be assumed that this is some magic finding that is not already included in good energy demand projections—though unfortunately many projections fail to factor past efficiency gains into future demand estimates. To slow supply growth beyond what can be done through efficiency requires voluntary and mandated demand reduction."[23]

Yet slowing energy growth "beyond what can be done through efficiency" is precisely what will happen if the Conserver Cult has its way. The cult of course declares that this is not the case, that energy conservation can be achieved with little or no effect on GNP, living standards or lifestyles. The Council on Environmental Quality, for example, states that "the United States can do well, indeed prosper, on much less energy than has been commonly supposed . . . the principal basis for this good news is the accumulating evidence that the means are available to wring far more consumer goods and services out of each unit of fuel that we use. . . . Energy productivity . . . thus refers to getting more from the energy we use, not to a back-to-the-caves reduction in amenities. . . . With a determined national effort to increase energy productivity . . . total U.S. energy need not increase greatly between now and the end of the century, perhaps by no more than 10–15 percent."[24] Of course, growth in energy consumption of 10–15 percent over two decades is equivalent to the standard Conserver Cult vision of the future in which energy use grows at a rate of 2 percent a year until 1985, following which there is no growth in use, or even a decline.

But, to the cult, conservation is not so much a way of improving energy efficiency as it is a way, in Mitchell's words, to "impose some kind of frugality or modify [the] life style of people."[25] This is nowhere more apparent than in the area of transportation in general and private cars in particular. Cultists are in fact never more vehement than when castigating people who drive cars. Schumacher, for example, attacks highly developed transport and communications systems because they make people *"footloose"* and result in "chaos." He simultaneously praises the past of several centuries ago when there was "structure" and "people were relatively immobile."[26] Lovins considers investments in car travel wasted and unnecessary, stating that "large public projects centered round the private car—our biggest consumer *ephemeral* [emphasis added]—offer an excellent example of resources that deserve a better fate."[27] Lovins also applauds what he calls an "important idea" of social critic Ivan Illich, namely, that "the average American man drives about 7500 miles a year in his car, but the total time it takes him to do this and to earn the money to finance it is about 1500 hours, which works out to five miles an hour, and we can walk nearly that fast."[28] Very important, indeed. Meanwhile, Friends of the Earth flatly declares that the

salient fact about transportation is that there's too much of it. We are much too mobile, too much attracted to where we aren't and too little appreciative of where we are. Mobility undreamed of before this century is not a God-given right. It is a luxury, and a luxury we grow increasingly unable to afford. . . . The auto is a centrifugal force, and our society is in danger of flying completely apart. Fortunately, though, resource scarcity is bringing the era of the auto to an end. We must reshape our cities and towns once more, fitting them to the man or woman on two feet instead of four wheels.[29]

And Ralph Nader left no doubt in anyone's mind as to what he thinks about cars when he said, "I'm in favor of zero automobile growth."[30]

But historian, urbanologist and former Hudson Institute analyst B. Bruce-Briggs points out that what is really going on here is an elitist class *War Against the Automobile*, the title of his book on the subject. "At its most vulgar level," Bruce-Briggs comments, "the anti-automobile crusade is simply the attempt to drive the other guy off the road, particularly when he is not as sensitive, educated, or prosperous as we are. That lower-class slob has some nerve jamming highways in his junky old Chevy with his wife in curlers and his squalling brats beating on the rear window. People like that belong in mass transportation, in subways or buses, not clogging our roads and slowing us down. . . . So to drive the rabble off our highways we must divine elegant justifications—that cars are unsafe, that they are polluting, that they are chewing up the landscape, etc., ad infinitum." Then, we pass federal legislation which mandates that "all Americans must drive the sort of cars preferred by the

275

'New Class'—purely utilitarian vehicles. . . . The great bulk of the American people must be denied the vulgar pleasures of powerful or aesthetically pleasing automobiles. A federal requirement for small cars also protects the interests of the prosperous in general—no longer will it be possible for the slobs to load up their family in big cars and jam up the national parks or recreation areas."

However, the most important tactic to reduce the automobility of the average American slob, Bruce-Briggs goes on to say, is to "make driving too expensive for him. One of the most unattractive aspects of contemporary American civilization, recognized by all 'enlightened,' 'concerned,' 'thinking' people, is that mass prosperity is a vile thing. It is shocking that ordinary working people have enough money to live almost as well as we do. Indeed, some of them are so crude that they cannot distinguish between their worldly goods and coarse pleasures and our superior preferences. Worse, we are constantly exposed to their vulgarity—there is no way we can escape the sight of a Chevrolet Monte Carlo or a Cadillac Eldorado. Clearly, the standard of living has to go down. We should push the cost of driving up."[31]

Following this elitist approach to "energy conservation," Congress passed and President Ford signed the Federal Energy Policy and Conservation Act (EPCA) of 1975, which requires that all new cars average 27.5 miles per gallon by 1985. But, as Bruce-Briggs points out, a "car with consumption of 27.5 miles per gallon must be a small car. Automobiles *can* be made somewhat smaller and lighter, with the same interior space and almost as comfortable a ride, but a good-sized, comfortable car can only be made to average 20 MPG—not 27.5. So by 1985, according to the new law, the average car made in

America is going to be the size of a subcompact Vega or Pinto." So what will millions of consumers find when they go down to the auto showrooms in 1985? They will discover, says Bruce-Briggs, that "there is nothing larger than what they regard as kiddie cars."[32]

Meanwhile, consumers will be charged the additional $80 billion which it will cost auto manufacturers to produce these "kiddie cars"—three times what it cost to put a man on the moon.[33] And at an average annual gasoline price of one dollar a gallon, the resulting higher car costs and prices will exceed gasoline savings by $2 billion.[34]

But consumers can still take heart. There are exemptions for some cars making it unnecessary for them to meet EPCA mileage standards. What cars? Well, the Rolls-Royce for one. And exemptions are also being considered for the Mercedes and other imported cars which, as the *Wall Street Journal* points out, are "favored by the wine and cheese Georgetown set."[35]

It should also be noted that there will be a change in lifestyle other than cramped car quarters due to forced energy conservation in the form of small autos. An increasing number of people are not going to have much of any kind of lifestyle because they are going to be killed or crippled as a result of small car accidents. The percentage of fatalities and serious injuries resulting from accidents increases as much as three times as the size or weight of cars decreases.[36]

However, for those people who think that a Rolls-Royce or Mercedes may be beyond their budget or who don't wish to increase risks to their lives riding in small cars, there is an alternative. They can ride bikes. That is the energy conservation advice of the Citizens' Advisory Committee on Environmental Quality, a select group of folks from the intellectual, social and entertainment world, which informs us in a publication called "Citizen Action Guide to Energy Conservation" that "bicycles are 28 times as energy-efficient as cars" and are "fun, healthy, convenient, and useful."[37] Well, bicycles are so much fun that 40,000 people were injured and 1,100 people were killed in 1973 in bicycle accidents—about 2 percent of total fatal and 3.5 percent of total reportable accidents. Furthermore, the problem was growing at the time at an increasing rate of 14 to 15 percent a year, with bicycle deaths increasing at a rate faster than deaths caused by the automobile or any other transportation mode. In addition, people on bicycles usually have to ride directly in the highway right of way, resulting in exposure to a variety of air pollutants which in high concentration have proven hazardous to health. Cyclists are also faced with a few other difficulties which deter many people from cycling to work, such as inclement weather conditions, inability to carry additional packages, and time costs, according to the Environmental Protection Agency. And, what do you know, in very bad weather, the cyclist's exposure could precipitate colds, EPA tells us. However, EPA adds that cycling in fact is possible in very cold weather since "the cyclist is exercising and providing some body heat. With very warm gloves, a hat, thermal clothes, and thermal shoes, cycling can be pleasant in the winter."[38] Thanks, EPA, we needed that.

But, if it is important to get people off the highways and into small cars and onto bikes, it is equally essential to keep them from moving into the countryside where all the finer folk live, right? So, under the EPCA of 1975, federal energy standards are also now being devised that will alter the "appearance, shape or inner workings" of every private home, office building, hospital, school and factory in the United States after February 1981. The standards, which are expected to be incorporated into building codes all over the country, will for the first time require that all buildings be designed to meet an "energy budget." Standards will be set lower than what would be possible under technically feasible performance but much higher than existing construction practices. And, of course, building and housing costs will be higher—if any new houses are built in the first place. Heating and air conditioning engineers are already maintaining that the performance standards developed so far are "unworkable."[39]

This "energy budget" law will be a wonderful, cost-raising, exclusionary tool in the hands of what Bernard J. Frieden, MIT professor of urban studies and planning, calls the "growth control and environmental coalition." Frieden relates that "when energy conservation became an important item on the national agenda, environmentalists and other growth opponents were quick to claim that suburban homebuilding would lead to wasteful energy use because of the low-density living pattern it creates. When a new study, *The Costs of Sprawl*, agreed with this position, they picked it up and used it uncritically. A close analysis of *The Costs of Sprawl* has shown that it greatly exaggerated the energy savings that might be achieved. . . . The energy conservation argument then is more an expression of doctrine than a result of analysis." Frieden also states that in California the cult, which includes local Sierra Club chapters, has "made a clear and substantial contribution to the escalation of new home prices; yet its success in discouraging homebuilding has failed to produce important environmental benefits for the public at large. Instead it has protected the environmental, social, and economic advantages of established suburban residents who live near land that could be used for new housing. . . . In one case environmentalists helped defeat a proposal that was favorable to both regional and local environmental considerations, apparently on the principle that the best development is no development at all."[40]

The shape of things to come may be emerging in Portland, Oregon where, seeking the carrot of federal hydropower allocations, officials are drafting what is described as the "most comprehensive energy conservation regulation in the United States." The proposed regulation includes mandatory "weatherization" within five years of all privately owned structures and "energy audits" for all homes and businesses. The cost to property owners for remodeling is estimated at $300 million, with each of 150,000 private home owners required to pay an average of $1,350 to install additional insulation, storm windows or other improvements required by the plan. If the improvements are not made after five years, the home owner would be barred from selling his house. Lower-income residents might be given financial assistance to pay for weatherization, but middle-income residents will have to pay for conservation improvements themselves and obtain loans at regular interest rates. All of this will reportedly provide a reduction in energy use of 30 percent and an energy-dollar savings of $160 million by 1995. But the cost of the energy conservation program will be $300 million—a net loss of $140 million on the project.[41]

But, having plunked down their $1,350 apiece for energy conservation in the form of mandatory "weatherization," etc., home owners will be happy to know that they will have substantially contributed to a new problem called "indoor air pollution," described by World Health Organization epidemologist Jan Stolwijk as probably doing more damage to human health than outdoor pollution. According to Tulane University Medical School internist Dr. George Burch, the problem is that by tightening up the American home "we may be transforming it into a virtual gas chamber," due to pollutants from household chemicals and other sources and naturally produced, radioactive radon gas trapped inside houses instead of being vented to the outside. In one model insulated home with advanced weatherproofing, in fact, a government researcher found that accumulations of radioactive radon gas were ten times higher than in a more typical, "leaky" house.[42] But there is no need to worry. All that has to be done is to promulgate another regulation requiring the home owner to install an air-purification system similar to those on submarines—at an additional cost of $5,000. But nobody ever said there was any such thing as a free lunch anyhow, did they?

Yet another and even more effective way of keeping people out of cars and suburban, single-family homes is to increase unemployment so that they don't have the income to afford such undeserved luxuries in the first place. This result of energy conservation is, of course, denied by promoters of reduced energy use. John Holdren, for example, states that "less energy can mean more employment. The energy-producing industries comprise the most capital-intensive and least labor-intensive major sector of the economy. Accordingly, each dollar of investment capital taken out of energy production and invested in something else, and each personal-consumption dollar saved by reduced energy use and spent elsewhere in the economy, will create more jobs than are lost."[43] Which is all fine. But what if there is no energy available to fuel "more jobs" after taking those dollars of investment capital out of energy production? . . .

Of course, we are moving more and more towards a service-oriented economy, and service activities in theory are less energy-intensive than industrial manufacturing operations. Yet, in the past, service activities have grown to account for about half of total economic output with no perceptible effect on the rate of increase in energy consumption. Furthermore, it is not at all clear that service activities are actually any less energy-intensive than industrial operations, for, as Schurr explains, services "are highly heterogeneous and some of them may turn out to be quite energy-intensive. For example, consider leisure activities, which in the future will account for an increasingly large percentage of the personal services consumers in advanced economies will demand. It is not unusual for people to travel great distances by airplane or automobile for a skiing weekend or to engage in other types of leisure activity that require substantial travel. This is, obviously, an energy-intensive form of service. There is also a growing trend in second homes—the future counterpart, perhaps, of the earlier phenomenon of second and third cars. The construction of such homes and the travel required to go from the city residence to the weekend house may both turn out to be comparatively energy-intensive. Thus, we should not fall into the trap of believing that the growth of nonindustrial activities in the future will necessarily be associated with lower intensities of energy use."[44]

But, in another sense, it certainly is true that turning off growth in energy production will increase employment or, more to the point, physical work done by people. This is because, if sources of inanimate energy are lacking, people will have to use their own human energy—the least efficient and most expensive energy source of all to do the work themselves. The Meinels give an example of what happens in this case:

> A large rock must be moved and there is $200 in the budget to accomplish the task. In Country A the contractor assembles 100 laborers and the rock is moved. A typical distribution is for the contractor and the laborers to share equally: $100 goes to the contractor and $100 to the laborers, with each laborer receiving $1. In Country B the contractor employs one machine and operator to move the rock. They also share equally: $100 to the contractor and $100 to the operator and machine. Thus, through the application of technology, Country B requires higher energy use than Country A, but gains a lower inequality of income.[45]

Or, to put it another way, the more a country uses human energy rather than inanimate energy to do physical work, the greater is the inequality of income between the rich and the poor. The Meinels, in fact, did a study which compared income inequality in different countries with their per capita energy use. They found that countries with lower per capita energy use show higher income inequalities between the richest 20 percent and the poorest 20 percent. "Energy use," they concluded, "is tied closely to energy costs and hence energy availability. A high price on energy will restrict its use and according to our economic modeling will force a higher inequality of income on the population. The poor will become poorer and the gap between the poor and rich larger."[46] Of course, but what else is new? Well, what else is new is that such a revelation would apparently not get Amory Lovins too upset because, as far as countries with low per capita energy use are concerned, he tells us that "we must not forget that in many countries, the main source of energy is people. People are extremely efficient users of food (if they get enough), and the food itself takes little energy to grow . . . if energy-intensive agriculture has not taken over from traditional methods. Those who believe that nothing substantial can be done without machines and heavy industry should examine the People's Republic of China."[47] Thanks, Amory, we needed that, too.

Well, even so, energy conservation at least will solve the "energy crisis," right? No. All that energy conservation ends up doing, as one engineer put it, is "delaying the time you are going to reach the end of a dead-end street."[48]

But energy conservation will still undercut the OPEC cartel, won't it? Nope. It will actually make the cartel stronger. The cartel is like the bully on the block. The more somebody gives in to him, the more he wants. The more energy conservation there is, the more people need the remaining energy and consequently the more they will be willing to pay for it. This will make it possible for OPEC to produce less oil, charge more for it and get away with it. Energy conservation thus ends up putting OPEC in an even stronger position because the cartel can make the same or even more profit through higher prices on less output, enabling it to keep even more of its oil "banked" underground for the future.

However, the really interesting thing about energy conservation is that it seems to be addictive. Once somebody figures out a way to save a little energy, it seems he can't stop trying to save even more and more and more energy. There was a time, for example, when some people thought we would need about 160 or more quads by the year 2000. Then this was scaled down to 120 quads. Following this, Lovins said we would only need 95 quads.[49] Now, there is another even lower-energy scenario produced at the University of Wisconsin which says that we can reduce energy consumption to about 33 quads in the year 2050.[50] To bring about this reduction in energy use, changes will have to occur in social and physical institutions and arrangements, according to the scenario authors. These changes will include new community patterns and shifts in the mix of goods, services, employment patterns, distribution of work, leisure and income, all of which will be in anticipation of "the end of the continued growth of the commodity component of society's output."

Space does not permit listing all of the "changes" envisioned by the scenario authors. But one does stand out as perhaps representative of their general tenor. This is the "phase-out of fast food services and other 'junk' commerce. . . . The popular fact that McDonald's hamburger chain uses enough energy a year—largely for packaging—to provide electricity for the cities of Pittsburgh, Boston, Washington, and San Francisco illustrates the American penchant for throwaway containers."[51] Of course, if they had asked the kids, the vote would have been unanimous to turn out the lights in Pittsburgh, Boston, Washington, and San Francisco. But who would have thought that Ronald McDonald was an enemy of the people?

Is it possible that Americans might object to the growing costs, increased health and safety problems, and expanding poverty resulting from energy conservation? Well, of course, any kind of insensitive behavior is possible from these people, but this has already been planned for too. The Consumer Motivation and Behavior Branch of the federal Energy Research and Development Administration has provided a study grant of $500,000 to a professor at the University of Denver to "test social incentives as a means of reducing energy consumption." But a "social" approach to American behavior control has already been devised, according to the *Mindszenty Report,* which quotes the August 1977 issue of *Human Behavior* magazine in which one Kenneth Lamott states that "President Carter has told us soberly that if we fail to change our ways, the alternative will be an unparalleled disaster . . . if we are going to survive in a livable world, we are going to have to learn to behave, both privately and publicly, differently than we do now." Lamott asserts that the notion that "Americans will voluntarily deny themselves anything is a generous delusion . . . if we are going to survive, we must change our behavior en masse by methods and techniques that are already at hand and that have been proved to work." He goes on to observe that "the late Mao Tse-tung was the greatest behavioral psychologist of the 20th century. Under Mao's guidance, a quarter of the world's people went through the greatest mass process of behavior modification in history. This great achievement was accomplished by psychological means." Behavioral scientists so far, Lamott laments, "haven't discovered how to steer us away from our greedy destruction of the environment. Maybe we ought to take a leaf from the textbook of that eminent Chinese psychologist, Chairman Mao. He had his ways, and they seemed to work pretty well. No matter how much their behavior may have been manipulated, the great bulk of Chinese are reasonably happy, secure in their conviction that they are part of a great historic movement that has raised up the poor and helpless."

While the ERDA-commissioned study should be interesting, Lamott continues, we can't wait for ERDA's results. "We must, as Carter has recognized," he says, "begin with the means we have at hand." Among Lamott's proposals, which he calls the "American Way," are the following:

• "The president will declare a state of emergency equal to a major war, following his declaration with symbolic actions equivalent to FDR's bank holiday."

• "The mass media will launch a vigorous campaign

designed to create a national spirit equal to the spirit of WWII, our last 'good' war."

● An American Way Administration will organize every man, woman and child into committees to monitor the use of energy in their offices, neighborhood schools and factories.

● These committees will meet regularly in self-criticism sessions with each member "trained in the dynamics of the encounter group, with particular attention to the hand-to-hand combat of the Synannon 'game.'"

● "Incorrigible wasters—probably about 10 percent of the population" will be expelled from their respective committees "and their names forwarded to the American Way Administration."

History has proven, sums up Lamott, that Mao's behavioral modification "prescription worked . . . as Yale psychiatrist Robert J. Lifton pointed out, the Chinese model was more powerful than George Orwell's because each person was monitored by many pairs of human eyes rather than by the impersonal telescreen of '1984.'"[52]

All of which, of course, seems kind of far out—until one remembers President Carter's Youth Energy Project. This was a plan concocted by his personal assistant, Greg Schneiders, in which the nation's youth were to be enlisted as energy detectives. Columnist Richard Spong tells how the youngsters were to "start outside a house, checking off energy problems on a government-prepared survey list. Then they would knock on the door to tell the occupant his or her outside energy conservation 'score' and offer to go through the inside of the house to complete the survey. Schneiders says the youngsters would use a 'team members manual' to check everything from tires to toilet tanks and the whole business shouldn't take more than 45 minutes to an hour." The project was to use a manual of nearly 100 pages, including not only the checklist but mottoes for conservation such as: "Be aware that mental and physical health probably will improve if individuals become more self-reliant and less dependent on energy-intensive life-styles." Spong reported that, according to Schneiders, the "plan is entirely voluntary, the information is not going to be sent anywhere, and nobody's going to be reported for anything. No Hitler Youth stuff." But, as Spong added parenthetically, the "scheme does, however, have a striking resemblance to the game plan for Mr. Fidel Castro's Pioneer Youth, who, when not otherwise engaged, make house-to-house bottle collections and neighborhood patrols to ferret out electricity-wasters."[53]

The Youth Energy Project never got off the ground. But the ideology which promoted it is alive and well and continuing to prosper among Conserver Cultists. It is the kind of ideology which establishes minimum cooling and maximum heating temperatures in commercial buildings throughout the country or, as University of Rochester political science professor William H. Riker puts it, is

used to justify the existence of a planned society which works, if it works at all, only in the context of a police state. Much of the ordinary workaday life would have to be made illegal to force society to make fuel conservation its highest priority. The electrician who wires a house for comfort rather than conservation, the salesman who speeds on his route by automobile rather than available public transit, the mayor who delays building a sewage plant because he cannot find money for it—all these will be transformed into commercial criminals. We could make "energy conservation a matter of highest national priority" . . . ; but the kind of life for which we conserved it would not be very attractive.[54]

15

THE COMING ENERGY CATASTROPHE

The game plan is simple. First, turn off growth in domestic energy production with environmental laws, safety regulations, and price controls. Then, when shortages develop, blame energy producers because they are not producing more energy, accuse them of "greediness" and "profiteering," and grab more political power to enforce greater control over and eventual nationalization of domestic energy production. Next, when shortages persist, blame energy consumers because they are using too much energy; accuse them of being wasteful energy pigs, and grab still more political power to enforce increasing energy conservation and growing control over people's lives. Finally, use this escalating power over energy producers and consumers to usher in the low-energy, low-technology, decentralized Conserver Cult vision of the future.

But what would life really be like under this future "soft" regime? Here, the picture gets a bit vague. We are presented only with beguiling images rather than detailed schemata of the new "soft life" ordained by the cult. E. F. Schumacher, for example, promises "health, beauty and permanence." But so do advertisers of facial creams. Amory Lovins tells us that the "most attractive political feature of soft technology and conservation . . . may be that, like motherhood, everyone is in favor of them."[1] Which is a curiously ironic choice of similes since Lovins also declares that "stabilization or reduction in population size" is one of the "changes in lifestyles" that "governments and their constituencies in rich countries should begin to contemplate seriously."[2]

Robin Clarke, editor of *Notes for the Future, An Alternative History of the Past Decade,* imagines the new, soft utopia looking "something like this: a countryside dotted with windmills and solar houses, studded with intensively but organically worked plots of land; food production systems dependent on the integration of many different species, with timber, fish, animals and plants playing mutually dependent roles, with wilderness areas plentifully available where perhaps even our vicious distinction between hunting and domestication was partially broken down; a life style for men and women which involved hard physical work but not over excessively long hours or in a tediously repetitive way; an architecture which sought to free men from external services and which brought them into contact with one another, rather than separated them into cubicles where the goggle box and bed were the only possible diversions; a political system so decentralized and small that individuals—all individuals—could play more than a formal, once-every-five-years role; a philosophy of change that viewed the micro-system as the operative unit; and a cityscape conceived on a human scale and as a centre for recreation."[3]

But perhaps the most comprehensive view of what Conserver Cult life might be like is provided in a work of social science fiction called *Ecotopia* by Ernest Callenbach. Published in 1975, *Ecotopia* was greeted with acclaim by Ralph Nader, who stated that "the book's impact . . . is the breadth of perspective that envelops the reader. None of the happy conditions in Ecotopia are beyond the technical or resource reach of our society." A reviewer for *Not Man Apart,* the Friends of the Earth newsletter, was even more effusive, gushing that "nearly every sentence offers yet another concretely sensible proposal, giving the book substantial density. . . . As a first primer for how to get it together as a stable-state, humanitarian society, it is the best book I've seen yet." The book was also described by something called *Sipapu* as "likely to be the underground classic of a generation . . . our Book of the Year."[4]

The theme of the book is set at the beginning by a quote from ecologist Barry Commoner: "In nature, no organic substance is synthesized unless there is provision for its degradation; recycling is enforced." Now, one might think that the title of the book, *Ecotopia,* stands for "ecological utopia." But this is not the case. According to the author, *Ecotopia* is derived from the Greek *oikos* meaning "household" or "home" and *topos* meaning "place." Set in the year 2000, the book is the story of one William Weston, a reporter who is sent by his New York newspaper as the first American in twenty years to visit Ecotopia, a country incorporating the states of Washington and Oregon and northern California, which seceded from the United States and declared its independence in 1980. Weston reports that

> it is widely believed among Americans that the Ecotopians have become a shiftless and lazy people. This was the natural conclusion after Independence when the Ecotopians adopted a 20-hour work week. Yet even so no one in America, I think, has yet fully grasped the immense break this represented with our way of life. . . . What was at stake, informed Ecotopians insist, was nothing less than the revision of the Protestant work ethic upon which America had been built. The consequences were plainly severe. In economic terms, Ecotopia was forced to isolate its economy from the competition of harder-working peoples. Serious dislocations plagued their industries for years. There was a drop in Gross National Product by more than a third. But the profoundest implications of the decreased work week were philosophical and ecological: mankind, the Ecotopians assumed, was not meant for production, as the 19th and early 20th centuries had believed. Instead humans were meant to take their modest place in a seamless, stable-state web of living organisms, disturbing that web as little as possible. This would mean sacrifice of present consumption, but it would ensure future survival—which became an almost religious objective, perhaps akin to earlier doctrines of "salvation." People were to be happy not to the extent they dominated their fellow creatures on the earth, but to the extent they lived in balance with them. The deadly novelty introduced into this accepted train of thought by a few Ecotopian militants was to spread the point of view that economic disaster was not identical with survival disaster for persons—and that, in particular, a financial panic could be turned to advantage if the new nation could be organized to devote its real resources of energy, knowledge, skills, and materials to the basic necessities of survival. If that were done, even a catastrophic decline in the GNP (which was, in their opinion, largely composed of wasteful activity anyway) might prove politically useful. In short, financial chaos was to be not endured but deliberately engineered. With the ensuing flight of capital, most factories, farms and other productive facilities would fall into Ecotopian hands like ripe plums.

Within several months after Ecotopia declared independence, pressure built up for the United States to intervene militarily, Weston relates. But the Ecotopians had already trained a nationwide militia and airlifted arms from France and Czechoslovakia. In addition, it was believed that at the time of secession they had mined major eastern cities with atomic weapons. Washington therefore decided against an invasion, setting off a wave of closures and forced sales of business in Ecotopia "reminiscent of what happened to the Japanese-Americans who were interned in World War II." The transition period that ensued was hectic, reports Weston, adding that

> certainly many citizens were deprived of hard-earned comforts they had been used to: their cars, their prepared and luxury foods, their habitual new clothes and appliances, their many efficient service industries. These disruptions were especially severe on middle-aged people—though one now elderly man told me that he had been a boy in Warsaw during World War II, had lived on rats and moldy potatoes, and found the Ecotopian experience relatively painless. To the young, the disruptions seem to have had a kind of wartime excitement—and indeed sacrifices may have been made more palatable by the fear of attack from the United States. It is said by some, however, that the orientation of the new government toward basic biological survival was a unifying and reassuring force.[5]

Taking actions that would be "impossible under the checks and balances of our kind of democracy,"[6] Ecotopia passed economic laws which "proliferated wildly in the obsessive attempt to shape all agricultural and industrial enterprise into stable-state, recycling forms.[7] There was a "massive flight of capital, similar to what happened after the Cuban revolution." People, seeing the former owners depart, realized that a "new era was indeed upon them and began spontaneously taking over farms, factories and stores." Laws formalizing the "forfeiture of property by owners, plus confiscatory inheritance taxes were legislated." Laws also redefined the position of employees. Workers in Ecotopian enterprises all became "partners" and the practical size of these "joint-ownership firms became less than 300 people." Meanwhile, massive and deliberate economic changes took place in the "diversion of money and manpower toward the construction of stable-state systems in agricultural and sewage practices, and in the scientific and technical deployment of a new plastics industry based upon natural-source, biodegradable plastics."[8]

Cars were eliminated in Ecotopia and replaced with electric trains, minibuses, bicycles, and walking. New, decentralized minicities were built in place of existing cities, which were razed and returned to grassland, forest, orchards, or gardens.[9] "Ecologically offensive" consumer items such as electric can openers, hair curlers, frying pans, and carving knives are "unknown." To curb industrial proliferation, many basic necessities are "utterly standardized." Ecotopians repair their own things and there is even a law which requires that pilot models of new devices must be given to a public panel of ten ordinary people. Only if all of these people find they can fix likely breakdowns with ordinary tools is "manufacture permitted."[10]

Ecotopia is organized on the basis of a stable-state concept that affects "every aspect of life, from the most personal to the most general."[11] Many Ecotopians are "sentimental about Indians" and "envy the Indians their lost natural place in the American wilderness." What matters most in Ecotopia is the "aspiration to live in balance with nature, 'walk lightly on the land,' treat the earth as mother." It is no surprise, comments reporter Weston, that "to such a morality most industrial processes, work schedules, and products are suspect! Who would use an earth-mover on his own mother?"[12] Ecotopians regard trees as "being alive in almost a human sense."[13] They have attitudes that can almost be called "tree worship."[14] At the same time, Ecotopians have a "secure sense of themselves as animals"[15] and believe human beings are "tribal animals."[16] Consequently, in many areas of Ecotopian life, there is a strong trend to "abandon the fruits of all modern technology, however innocuous they may be made, in favor of a poetic but costly return to what the extremists see as 'nature.'"[17]

Ecotopia is marked by rapidly disappearing nuclear families, communal groups of between five and twenty people, and a life which is "strikingly equalitarian in general." Ecotopians are remarkably healthy-looking, used to walking everywhere carrying heavy burdens like backpacks and groceries for long distances, and have a "generally higher level of physical activity than Americans."[18] People are open and friendly in Ecotopia, gathering round and talking, according to Weston, the "kind of leisurely talk I associate with college days. Jumps around from topic to topic, and people kid a lot, and cheer each other up when need be."[19] Everybody gets a "kick out of the expression of intense feeling," he remarks, adding that "evidently restraints on interpersonal behavior have been very much relaxed here," and even "extreme hostility can be accepted as normal behavior."[20]

After secession, Ecotopia adopted a formal national goal of a declining population. Population has tended to drop at a rate of around 65,000 per year, so that the original Ecotopian population of some 15 million declined to about 14 million (nothing is mentioned about people emigrating or being allowed to emigrate from Ecotopia aside from rich families whose property was expropriated). However, according to Weston, this is not enough for some radical thinkers of the ruling Survivalist Party, who believe that a proper population size would be "the number of Indians who inhabited the territory before the Spaniards and Americans came—something less than a million for the whole country, living entirely in thinly scattered bands!" However, Weston points out that "most Ecotopians contend that the problem is no longer numbers as such. They place their faith for improvement of living conditions in the further reorganization of their cities into constellations of minicities, and in a continued dispersion into the countryside."[21]

Ecotopian energy production moved uniformly toward power sources like solar energy, earth heat, tides, and wind, which supposedly can be tapped indefinitely without affecting even the local biosphere. Ecotopians thus take a childish delight, Weston says, in the windmills and rooftop wind-driven generators that are common in both cities and remote areas. The Ecotopians inherited a system of oil- and gas-fired power plants, which they closed within a few years, and a number of atomic fission plants. They have been willing to "live temporarily with the fission plants located in removed and little-inhabited areas—though they have redoubled engineering precautions against nuclear explosions and extended hot-water discharge pipes more than a mile to sea." According to Weston, energy in Ecotopia is inordinately expensive, costing about three times what it costs Americans. One curious symptom of the high cost of energy, he says, is that "houses tend to be abominably ill-lit. They contain lamps of several kinds, used for reading and work purposes—though Ecotopians avoid fluorescent tubes, claiming their discontinuous emission patterns and subliminal flicker do not suit the human eye. But for ordinary socializing their houses are lit by small bulbs and often even by candles (which they produce from animal fats as our ancestors did)."[22] Many city dwellings are heated by a system using solar radiation in a large water tank underground, from which heated water can be pumped through radiators in the living areas. The Ecotopians also have a massive thermal gradient [OTEC] power plant and many similar, smaller plants along the coastline. In addition, they are working on a process whereby, in specially bred plants, photosynthetic chemistry would be electrically tapped directly. "Such an unbelievably elegant system," points out Weston, "would be nearly perfect from an Ecotopian point of view: your garden could then recycle your sewage and garbage, provide your food, and also light your house!"[23]

Weston meets Vera Allwen, the president of Ecotopia, whom he describes as "a very direct person. Despite being rather small and a trifle stout, gives off a strong air of authority. Clearly well used to exercising power. But not businesslike and cool about it, like many of our politicians, who are sometimes hard to tell from businessmen—heads full of impersonal calculations that happen to be power equations instead of money ones. She is powerful as a person, *not* as a bureaucrat or the head of an institution. Difficult to express. (Have heard that some of the old-time communist leaders, Ho Chi Minh and Mao Tse-Tung, had this quality too.)"[24]

There is more, much, much more, e.g., houses made from plastic extruded from cotton so they can be biodegraded and recycled;[25] unarmed police;[26] legalization of marijuana and other drugs;[27] elimination of

297

"victimless" crimes such as prostitution, gambling, and drug use;[28] punishment of deliberate pollution of water or air by severe jail sentences;[29] etc. But suffice it to say that Weston falls in love with an Ecotopian girl named Marissa, achieves a "breakthrough" in which he suddenly hears his voice saying "I am going to stay in Ecotopia" while floating in a warm bath up to his nose, and pens a note to his New York editor informing him that "I've decided not to come back, Max. . . . But thank you for sending me on this assignment, when neither you nor I knew where it might lead. It led me home."[30]

The *Oregon Times* described *Ecotopia* as "not your standard Utopia; in fact it appears to be the first *pragmatic* utopia ever written. And in that sense it is the first distinctively American utopia."[31]

Whether or not *Ecotopia* is a "standard utopia" much less a "pragmatic" or "American" one, it is clear that it is a standard totalitarian dictatorship, except that control is exercised not by political masters representing the State but ecological ayatollahs serving Nature. *Ecotopia* makes clearer what Ralph Nader meant when, as H. Peter Metzger relates, he stated that we will reduce our energy consumption by half, "but first there has to be a intermediate stage of liberalism failing, and of economic bad times without novocaine."[32] It also makes clearer why Debbie Galant, editor of *Environmental Action* magazine published by Friends of the Earth, wrote a rebuke of the FOE book called *Progress as if Survival Mattered,* which depicts the environmentalist organization's utopian Conserver Society of the future. In a review, Galant wrote that it "promises, in 320 difficult pages, to provide 'unfrightening, tempting' steps toward the utopia of a 'Conserver Society.' I believe in the Conserver Society that Friends of the Earth is after. But the steps outlined in

Progress as if Survival Mattered are not at all unfrightening or tempting to me. FOE's solutions hinge mostly on more rules, more laws, more government—which comes down in the end to more restrictions on individuals." Galant quoted a statement of FOE president David Brower that "coercion by many governments will undoubtedly be required [to control population growth]." Said Galant: "I can't wait." Brower presumes, she wrote, that "everybody on earth already agrees with his viewpoint on how to change the world. This decision-making process seemed a little undemocratic to me, considering that Friends of the Earth's staff and membership is probably less than a millionth of one percent of the world's population." She concluded that "I'm waiting for the sequel, *Survival as if Dignity Mattered.*"[33] But Galant will apparently not wait as a member of the FOE staff because, at last report, she was no longer an employee of the environmentalist organization.

However, what would living really be like under the conditions described in *Ecotopia?* In Great Britain's Wales, John Seymour and his wife, Sally, spent eighteen years working towards a personal utopia which Seymour describes as "post-industrial self-sufficiency: that of the person who has gone through the big-city-industrial way of life and who has advanced beyond it and wants something better." But Seymour spares no sacred cows when he relates that

for the past eighteen years Sally and I have been probably as nearly self-supporting with food as any family in north-west Europe. We have a very good idea of what it is like and what it involves, and therefore I feel qualified here to utter a solemn warning. It is beyond the capabilities of any couple, comfortably, to try to do what we have attempted. If a married couple settled down on five or ten acres of good land, in the British

299

climate, and devoted their entire time to being self-supporting in food, clothes and artifacts; and if they knew how to do it, and had the necessary stock and equipment, already paid for, they could succeed. They would be working just the fifteen hours a day, three hundred and sixty-five days of the year, that is, if they were to maintain the standard of living, and variety of food and of living, that they could maintain in a town. They would be very *healthy* doing this, they would not be *bored* (because they would never be doing the same job for long and would be doing a great variety of tasks), but they might sometimes wish they could sit down. Thoreau, when he lived at Walden, and wrote his famous book about it, lived almost exclusively on beans, and he didn't work very hard at all. He spent a very large part of his time there wandering around in the woods, peering into the depths of his pool, thinking and dreaming and meditating. I think he was a very sensible and enviable young man indeed. But he didn't have a wife and a family to bring up. Personally, I would not be prepared to live for two years and two months (which is the time Thoreau spent at Walden) on beans. Sally certainly wouldn't either, and we would be very hard put to make the children do it.[34]

Ecotopia, in fact, has no relationship at all to a real world in which conventional energy resources are turned off and reliance is basically placed on energy conservation and soft energy sources. What happens in the real world when energy is turned off is what happened, for example, when there were gasoline shortages in the summer of 1979. As lines outside filling stations got longer and longer, tempers got shorter and shorter. The most extreme case was in Levittown, Pennsylvania, where two nights of rioting over lack of gasoline left more than 130 persons injured and 200 were arrested.[35]

However, incidents like this are the least of it because following the Conserver Cult to its soft-energy vision of the future is, as Peter Brennan puts it, a "blueprint for disaster."[36] A study by Data Resources, Inc., for example, indicated that unless increasing energy needs are met, unemployment will be up, inflation will be galloping at double-digit rates, and business expansion will grind to a near-halt.[37] Another study by Milton Copulos of the Heritage Foundation found that deferral of further commitments to nuclear power could result in "significant economic and social dislocation. A shortage of over eight million jobs at an annual cost of more than $98 billion could develop. Blacks, women and other groups just beginning to make economic headway would see recent gains disappear. Most importantly, attempts to cope with large-scale shortages of electrical generation capacity could result in profound changes in the American lifestyle."[38]

H. Peter Metzger points out that "if environmentalists succeed in stopping nuclear power and in squeeze-play style also slow strip-mining and coal-fired plant construction, there will, of course, be serious nationwide power shortages. Rationing and part-time closings of industrial plants will follow and surely will result in widespread unemployment. By that time, it's likely that an awakened public demand will override environmental extremism and construction of nuclear plants will begin again. But there's a lag-time of six to 10 years between the decision to build a plant and the time it finally comes on line. It's during that period that social disruptions caused by massive unemployment would increase even though social decision makers might be doing their best to develop nuclear energy to get people back to work."[39]

Even more significant, turning off growth in production of conventional domestic energy will place greater demands on imported oil from Saudi Arabia and other nations. But, as *U.S. News & World Report*'s Dennis Mullin points out, it's "important to understand just how incredibly vulnerable Saudi oil is. The eastern part of Saudi Arabia produces most of the country's oil, and that is where the major reserves are located. There are only two power plants in that part of Saudi Arabia, and experts have told me that a small guerilla force in jeeps could blow them up. That would stop oil production for a year. An American Air Force squadron or an aircraft carrier couldn't stop such a raid."[40] More to the point, Fred Hoyle notes that, if "you were Russian, you would surely take careful note of that great crescent containing nearly 70 percent of world oil reserves which starts in the U.S.S.R. and sweeps through the Middle East into North Africa. . . . You would also notice the great bulk of Africa around which tankers from Europe and North America must go to reach the oil fields of the Middle East, and you would realize that control of the western coastline of Africa would permit you to cut those tenuous shipping lanes. So . . . to develop your muscle you would expand your navy, especially its submarine complement. You would also set yourself to exploit the many political troubles to which the continent of Africa is endemic. Believing in the all-importance of energy, you would scent victory in the world struggle."[41]

David Bodansky of the University of Washington's Department of Physics puts it even more bluntly, declaring that

if we abandon nuclear power for solar power, we may be writing a prescription for disaster, both economically and politically. In a world in which oil is becoming a scarcer and scarcer commodity, one cannot dismiss the possibility of stumbling into a war over oil, if the promise of solar power is slow to realization. The warning from Chancellor Helmut Schmidt of West Germany is pertinent: ". . . I will point to the great danger that if nuclear energy is not developed fast enough, wars may be possible for the single reason of competition for oil and natural gas." Of course, this may be alarmist. Nevertheless, we should not forget that Hiroshima came in a war which stemmed in significant measure from competition for natural resources in Southeast Asia.[42]

16

WHAT FUELS THE CONSERVER CULT?

The Conserver Cult is fueled by a diverse range of social, economic and political forces. But all of them point toward the revolutionary objective of seizing power through control of energy. As *Energy Daily* editor Llewellyn King puts it, energy "has been the sleeper of the industrial revolution, the true substitute of mechanical labor for physical labor. Because of this total penetration, it also represents a unique tool by which society can be shaped and controlled; a device by which the standard of living and the very nature of the national lifestyle can be controlled, modified, amplified or reduced. Since 1973, the link between energy and social engineering has been clearly exposed and has attracted those who seek political and social change; those who find today's society repugnant ... have seen manipulation of energy supply as a means of bringing about the restructuring of society to have it conform to their concept of a more utopian world. In short, it is the new highway of the revolution. . . ."[1]

This revolutionary power grab requires turning off energy production and consumption regardless of the catastrophic consequences for the American people. But, from the Conserver Cult perspective, the only catastrophe worth talking about is a situation in which there is *too much,* not too little, energy. Lovins expresses this view when he says that "if you ask me, it'd be little short of disastrous for us to discover a source of clean, cheap, abundant energy *because of what we would do with it.* We ought to be looking for energy sources that are adequate to our needs, but that won't give us the excesses of concentrated energy with which we could do mischief to the earth or to each other."[2]

However, the Conserver Cult does not speak of energy catastrophes but rather of a beautiful, new, decentralized future in which we will disconnect from the centralized power facilities of the present and plug into the local, renewable resources of the future. This will make it possible, according to Lovins, for our "lifeline" to come from "an understandable neighborhood technology run by people you know who are at your own social level" rather than an "alien, remote, and perhaps humiliatingly uncontrollable technology run by a faraway, bureaucratized, technical elite who have probably never heard of you."[3] But, as engineer Daniel W. Kane points out, this theory is "reminiscent of certain ideas utilized by the People's Republic of China during the years of the 'Great Leap Forward.' At that time, it was stated by certain Chinese policy makers that new large 'central station' iron and steel production plants were not needed to double the Chinese steel production—what was needed were thousands of backyard iron furnaces and steel converters. The

Chinese proceeded to build their backyard iron furnaces and steel converters by the tens of thousands. It is rumored that as many as 20 million Chinese may have been involved in the effort. The backyard steel turned out to be unusable due to its poor quality and the 'Great Leap Forward' became the 'Great Leap Backward.' "⁴ In other words, stripped of their cars and other energy-wasteful "ephemerals" and relocated in decentralized communes in the countryside, Americans might find that the "lifeline" coming from "an understandable technology run by people you know" consists of a generator connected to a bicycle which they pump at the behest of people they know as local energy commissars.

This comparison with the People's Republic of China is apropos because Conserver Cultists express an inordinate admiration if not adulation for the late Mao Tse-Tung and his methods. Schumacher comments, for example, that "many of us have been struggling for years with the problems presented by large-scale organization, problems which are becoming ever more acute. To struggle more successfully, we need a theory, built up from principles. But from where do the principles come? They come from observation and practical understanding. The best formulation of the necessary interplay of theory and practice, that I know of, comes from Mao Tse-Tung. Go to the practical people, he says, and learn from them: then synthesize their experience into principles and theories; and then return to the practical people and call upon them to put these principles and methods into practice so as to solve their problems and achieve freedom and happiness."⁵ And Ralph Nader observes disparagingly that "compared to China, we can't get anything done. China eradicates VD; we can't even stop it from increasing. Is it because we don't have the communications system, the

306

technology, the laws? They have much more serious communication and transportation obstacles than we do—they don't have a telephone system or a television system with a unit in every house. But we don't have the coordinated determination of marshaling the massive numbers of people in a common cause."[6]

To develop this "coordinated determination," the Conserver Cult first of all fuels itself on what British historian-author and former *New Statesman* editor Paul Johnson calls the "Ecological Panic" in his book *Enemies of Society*. Once ecology became a fashionable, good cause as it did in the late 1960s, Johnson points out, reason, logic, and proportion

flew out of the window. It became a campaign not against pollution, but against growth itself, and especially against free enterprise growth—totalitarian Communist growth was somehow less morally offensive. . . . One of the most important developments of our time (I would argue) is the growth, as a consequence of the rapid decline of Christianity, of irrationalist substitutes for it. These are not necessarily religious, or even quasi-religious. Often they are pseudoscientific in form, as for instance the weird philosophy of the late Teilhard de Chardin. The ecology panic is another example. It is akin to the salvation panic of 16th century Calvinism. As I say in my book, when you expel the priest, you do not inaugurate the age of reason—you get the witchdoctor. But whereas Calvinist salvation panic may have contributed to the rise of capitalism, the ecology panic could be the death of it.[7]

Johnson goes on to say that

it is significant that the ecological lobby is now striving desperately with fanatic vigor and persistence, to prevent the development of nuclear energy, allegedly on the grounds of safety. . . . In Britain recently, we had a long, public enquiry, into whether or not it was right to go ahead with the enriched uranium plant at Windscale. The enquiry was a model of its kind. The ecolobby marshalled all the scientific experts and evidence they could lay their hands on. At the end the verdict was that there was no reason whatever why the program should not proceed. Did the ecolobby accept the verdict? On the contrary. They immediately organized a mass demonstration, and are planning various legal and illegal activities to halt the program by force. Now it is notable that a leading figure in this campaign is the man who is perhaps Britain's leading Communist trade unionist, Mr. Arthur Scargill of the Mineworkers. He has never, so far as we know, campaigned against Soviet nuclear programs, peaceful or *otherwise*. But the mass of the movement, in the U.S., Britain, France, Germany, and Italy, so far as I have been able to observe, is not politically motivated. They are simply irrational; but irrationality is an enemy of civilized society, and it can be, and is being exploited by the politically interested.[8]

The Conserver Cult is also fueled by the aversion of elite, upper-class members of society to anything which might affect their privileged positions, including advancement of people lower than themselves on the socio-economic ladder. "Most of the current no-growth advocates argue for a redistribution of resources as opposed to continued growth as a means of improving the current quality of life," says Herman Kahn, head of the Hudson Institute. "But we have argued that many of these 'reformers' do not mean what they say. They already have a high standard of living and do not see any real future gain for themselves if others improve their economic standards—although they may not recognize these as their true feelings."[9] Margaret Maxey observes that "conservation happens to be an option for, and a recommendation made by, the 'haves' in our society—those who can trim off the fat in patterns of consumption. It is a laudable program for the affluent and for the middle class. However, no such clamoring for 'doing without energy' is heard from blacks and other minorities, or from the poor, the elderly, the retirees, the people for whom self-denial is a steady, grinding, daily way of life." She quotes George Will to the effect that "our society is democratic, yet its social structure, as Mr. Will reminds us, is much more like the Titanic. When the Titanic rammed into an iceberg, the consequences of that disaster were far from democratic. Of 144 first-class women passengers, only four were killed. Out of a total of 325, 203 first-class passengers survived—62.5 percent. All the others who died—totalling 1,490—were not in first class. It is a fact of life that those whose lives are lived at or below the waterline in any society are harmed first and worst when the basic necessities of life are denied them."[10] And Andrew Cherlin, assistant professor of social relations at the Johns Hopkins University, writes that "perhaps the emphasis on personal concern is the reason that the antinuclear movement has caught on in the Seventies. At a time when people are turning inward, nuclear power arouses a middle-class constituency that is anxious about protecting its standard of living. . . . If this is the 'me decade,' as the writer Tom Wolfe has suggested, then nuclear power is the 'me issue.'"[11]

Making a comparison to the "leisure class" analyzed by Thorstein Veblen in his famous classic, *Harper's* magazine contributing editor William Tucker notes that the "environmentalists knew the language of energy and ecology, and could describe a future filled with windmills and with bright sunshine radiating 'inexhaustible energy.' Yet one never got the impression that these people were planning to be part of it. The 'soft energy' of the future was a vision offered to persuade people to forego the nastier, more vulgar realities of the 'hard energies' of the present." The message here is clear, Tucker continues, pointing out that "nuclear energy is 'hard' and 'dirty' and involves nasty realities of life. Solar energy, on the other hand, is 'soft' and 'clean.' It is most notable for the *lack of effort* it promises. The future will hold no more grubby realities such as digging coal out of the ground or drilling for oil—no more handling of dangerous materials. There will be nothing to do but sit back and watch the windmills revolve and the sun shine. The correct word for the environmental vision is not *clean* or *soft*. It is *genteel*. It is also a vision that will call us to disaster." However, as Tucker adds, the "leisure-class environmentalists will be perfectly content to leave things the way they are, regardless of the economic consequences, since, as Veblen notes, 'at any given time this class is in a privileged position, and any departure from the existing order may be expected to work to the detriment of the class rather than the reverse'.... The great appeal environmental solutions offer is that they can be worn like a badge of success. To say that one is an 'environmentalist,' or that one favors 'no-growth,' is to say that one has achieved enough well-being from the present system and that one is now content to let it remain as it is—or even retrogress a little—because one's material comfort under the present system has been more or less assured."[12]

It is most often this "leisure class" elite which forms the backbone of environmentalist and other organizations driving the Conserver Cult. Allan May, author of *A Voice in the Wilderness,* describes how environmentalist organizations such as wilderness conservation groups become extremist because they

cannot change their goals and tactics. They cannot change because they are largely controlled by profit seekers and pressure relievers whose interests are served by maintaining the program.

To change would cause dissension in the ranks, reduction in membership and finances, organizational problems, loss of power and prestige. None of those results would be desirable. . . . So they use the group's internal propaganda apparatus to keep the membership whipped up and the goals unchanged, regardless of the needs of the community. . . . The wilderness-conservationists this book opposes are probably members or officers of one or more of the more militant, uncompromising, close-minded environmental organizations such as but not limited to the Sierra Club, the Wilderness Society and Friends of the Earth. They are usually well-educated professional and semi-professional people who well know the workings of government and the means for influencing it. Sometimes they are the drop-outs whose goal often is to punish (or destroy) society for being society. . . . They form a clique, an establishment, an 'in' group, a mutual admiration society whose first requirement of member-ship is unquestioning acceptance of the credo. Members must also accept uncritically the dogma that the clique is made up entirely of superior people. Each individual thus feeds and at the same time feeds on the egos of all the others. They share an ability to reject or shut out external ideas. They often are ruthless, even vicious, and are convinced that the end justifies the means. Their actions are single-minded. They present questions and programs in either-or language. They seldom recognize trade-offs. They brook no compromise. They recognize no alternatives. They demand their goals in entirety. They condemn all other goals and the people who propose them. . . . The profit seekers use the crusade to further their own interests, to provide themselves with jobs or progressional aggrandizement or busi-ness opportunities. The pressure relievers use it to relieve their internal pressures. Some have psychological needs to harm people. . . . They take part in the battle not because they care about the goals but because it gives them a platform from which they may attack people without fear of being hurt themselves.[13]

Another force fueling the Conserver Cult is what is called the "new class." Michael Novak, Ledden-Watson Distinguished Professor of Religion at Syracuse Univer-sity and a resident scholar at the American Enterprise

311

Institute, points out that since World War II a "new social class has emerged whose main business is not business. The fact that there is such a class—the intellectual class, or more exactly, the class whose power base lies in 'the knowledge industry' and in the State—is not new. But two powerful changes have recently raised the status of that class: 1) it has grown enormously in numbers, both in its leadership cadres and in the millions of citizens whose cause is linked with theirs; and 2) simultaneously, powerful instruments of social change have emerged which are perfectly suited to its own needs and purposes. These are the national media of communication, especially television and radio, but also the national news magazines and the major national organs of daily news."

As Novak relates, a "new class employs new instruments of power, manifests new interests, exemplifies new ambitions, and attempts to make the whole culture reflect its own tastes, ideals, and styles of living. All four of these characteristics are manifested by this new class. We have witnessed the growth of the 'new politics' and 'media politics'; a rising interest in 'change,' that is, the displacement of older centers of power; new types of self-aggrandizement through new legislation and new forms of litigation (since the new class often prefers to acquire power through the courts rather than through the legislatures); and a 'new morality.' The new elites have chosen as their center of power the communications industry—the New York–Los Angeles axis—and their main base of operations the suburban communities near major universities (such as the Boston suburbs, Madison, Evanston, Ann Arbor, and Berkeley). Some heroes of the new elites have been Dan Rather, Robert Redford, Barry Commoner, George McGovern, Tom Hayden, Andrew Young, Gloria Steinem, Hunter S. Thompson, Ralph Nader, and John Gardner."

What distinguishes the left wing of this new class, Novak goes on to say, is "its hidden agenda of self-aggrandizement, its adversarial posture against the central conceptions of our political, economic, and cultural systems, and its attempts to short-circuit the will of the majority. . . . The new class, so defined, is the carrier of the new politics, the new ideology, and the new morality. . . . The new class specializes in 'communications.' The growing power of communications—within industry and between industry and the larger public—has brought about an expanded and powerful role for the new class even within industry, but within virtually every other segment of society as well. Far from being marginal, the new class now finds itself at the center of power. . . . If seizing effective control of the bureaucracies of government is seen to be the latent, and often manifest, intention of the new class, it has a very large reservoir of political power to be mobilized in the millions of government workers; professors, teachers, and staff of educational institutions; social workers; and workers in those industries whose economic base depends upon large government spending."[14]

Another perspective on this new class is provided by Irving Kristol, resident scholar of the American Enterprise Institute and coeditor of *The Public Interest,* who describes them as "people who are eager, as many among us are, to use the regulatory mechanism as a lever for transforming the social and economic order as a whole. I have in mind the people—the 'New Class' as they have come to be called—who do not, on balance, like a free, commercial society: that is, a society shaped by voluntary commercial transactions among consenting adults, in short, one whose central economic institution is the marketplace. They sincerely believe that a powerful government, in which they have positions of authority, can order things better. If one designates them, in a shorthand way, as 'Naderites,' one would not be far off the mark. . . . The plain fact is that these are people who are discontented—often in a vague and unfocused fashion—with modern technological civilization. They are inclined to believe that a 'planned' economic system would create a superior way of life for all Americans. They detest the individualism so characteristic of a free society, which is why they speak so favorably of 'mass transit' and 'planned communities' and so contemptuously of the private automobile and 'suburban sprawl.' They also, therefore, have little use for private economic enterprise and are utterly indifferent to the costs they load upon it."[15]

California's Sen. S. I. Hayakawa is even more blunt, stating that "Washington is full of power-hungry mandarins and bureaucrats who distrust abundance, which gives people freedom, and who love scarcity and 'zero growth,' which give them the power to assign, allocate, and control. If they ever win out, heaven help us! Americans do not know how to live with scarcity."[16]

One example of how the communications-oriented new class operates is provided by *Chicago Tribune* columnist Mike Lavelle, who relates that "David Rossin is an engineer who works for Commonwealth Edison [Chicago's electric utility]. Rossin was invited to a preview of *The China Syndrome* and he tells what happened: 'Jane Fonda, Michael Douglas, and Jack Lemmon had invited about 50 editors from college newspapers to the preview, which was to be followed by a question-and-answer session. A young girl asked a question critical of the movie and Michael Douglas asked her sneeringly if she worked for Commonwealth Edison. She said she had

314

once, but only part-time. Douglas asked if there was anybody else in the audience who worked for Commonwealth Edison. I raised my hand and then a guy came over to me and told me that I had to leave. I did not wish to make a scene, so I just left.'" What is disturbing about this propaganda blitz, comments Lavelle, is "the fanaticism that does not allow the other side a fair hearing. . . . With such a one-sided Q&A session it will be no surprise when our college newspapers reflect the totalitarian-type brainwashing in evidence. Whatever else Douglas believes, he does not believe in the give and take of democracy."[17]

The Conserver Cult is also fueled by militant, activist organizations which conduct sit-ins, occupations, blockades, and other "nonviolent" operations. One of the most prominent of these organizations is the Clamshell Alliance in New England, which has staged demonstrations and sit-ins at the site of a nuclear power plant being built in Seabrook, New Hampshire, since 1976. Since then, other groups such as the Shad Alliance in New York, the Potomac Alliance in the Washington, D.C., area, and California's Abalone Alliance, have become active.[18] The Clamshell and the other alliances are basically antinuclear, declaring that "with changes in the regulatory and political climate, renewable sources of energy—such as solar technologies—would become competitive, conservation would flourish, and the alleged 'need' for nuclear energy would vanish. Awareness of the fact that we live within a balanced, natural ecosystem necessitates changes in 'traditional' economic and social values."[19]

As the Conserver Cult has expanded its power, organizations such as the Clamshell Alliance have become increasingly more militant. In 1977, for example, Georgia Cong. Larry McDonald reported in the *Congressional Record* that one of the leadership roles in the Clamshell

Alliance had been taken by Samuel Holden Lovejoy of NOPE (Nuclear Objectors for a Pure Environment of Montague, Mass.). Since 1974, stated McDonald, "Sam Lovejoy has been the foremost advocate of sabotage to prevent construction or operation of nuclear fueled electric power plants. On February 22, 1974, Sam Lovejoy put his principals [*sic*] of support for sabotage into practice when he caused the collapse of a $50,000 preliminary weather monitoring tower at the Montague nuclear powersite. Lovejoy then surrendered himself to police and in a prepared statement to the press admitted 'full responsibility for sabotaging that outrageous symbol of a future nuclear powerplant.' After a week-long trial in which Lovejoy defended himself as having acted 'in the public interest,' and in which he was supported by a number of anti-Vietnam activists serving as character witnesses, the judge directed acquittal on grounds of a faulty indictment which charged Lovejoy with destruction of personal, rather than real, property. Lovejoy then triumphantly told the press, "The publicity . . . was a great victory, and we've entered the issue of civil disobedience into the environmental movement."

McDonald went on to point out that

since 1974, Sam Lovejoy has worked to popularize the concept of destruction of utility company private property as "non-violent direct action." Lovejoy has worked to forge alliances with activist groups such as the War Resisters League—WRL; Women Strike for Peace—WSP, a group thoroughly penetrated by the Communist Party, U.S.A., and which is seeking to combine its campaign for U.S. disarmament with opposition to the use of nuclear technology for peaceful purposes; and Ralph Nader's Critical Mass organization. Lovejoy had attended Nader's Critical Mass conference and expounded the need for sabotage to the cheers of the assembled activists. . . . Now Ralph Nader has become involved in supporting the concept of destruction of property as "civil disobedience." The Village Voice of April 4, 1977, published an interview with "public citizen" Nader in which he was quoted as follows:

"What activists are trying to do is make new law based on the settled Anglo Saxon tradition of self-defense that stretches back through Blackburn's commentaries," Nader replied. "That is, if someone tries to break into your house you can retaliate lawfully. In the case of a nuclear reactor, the self-defense is projective. But what are you going to do, wait until radioactivity is all over the place? Shouldn't you destroy property before it destroys you? Here you are violating a minor law to get judgment on a more important one, the way they did in the civil-rights movement when they sat at those lunch counters. You know," he said, gesturing sharply out the window of his office, "if it hadn't been for those demonstrators, the war would still be going on. The government was afraid of civil war. I'll make a prediction: if they don't close these reactors down, we'll have civil war within five years. . . ."

But what Nader is doing, protested McDonald, is "perverting our legal traditions which say that force may be used in self-defense against an immediate threat to your life. Our tradition does not say you can go out and kill your neighbor on the grounds that there is a statistical one-in-a-million chance that someday he might go crazy and kill you. The probability that all the multiple safeguards built into nuclear reactors would all fail and that significant amounts of radioactivity would be released is infinitesimally small. Yet on that miniscule probability, the anti-nuclear technology activists are trying to justify sabotage and mayhem."[20]

Another militant organization purportedly opposing nuclear power was formed in 1977, according to the Rockford College Institute, which reported that a "group of anti-war activists has organized and orchestrated a movement to arouse public resistance to nuclear power, but the group's purposes include only incidentally the possible dangers to the public inherent in the present state of nuclear technology. However, since a potential danger to the citizens lends itself to exploitation by those who wish to arouse public passions, anti-nuclear power was chosen as the umbrella issue under which to forge a coalition of militant groups among which the principal common denominator seems to be an aversion to capitalism." Named Mobilization for Survival, the coalition issued a "Call to Action," whose message began: "The proliferation of nuclear weaponry . . . and the increasing commitment of governments and private corporations to nuclear energy . . . is the central gravest threat to the human race today." The message asserted that funds which might otherwise have provided "adequate health care, a good education, decent housing, enough to eat, liveable cities and to help create more jobs and reduce inflation" were being spent for weapons wherein "only the defense contractors profit." Staging protests and demonstrations at nuclear sites throughout the country, Mobilization for Survival operates regional centers in San Francisco, Washington, Chicago, Boston, and Atlanta and fields teams of speakers, one of which featured Dr. Benjamin Spock, Daniel (Pentagon Papers) Ellsberg, environmentalist Barry Commoner, and socialist author Sidney Lens.[21]

Fueled by a diverse number of socioeconomic and political forces and embodied in a wide range of different organizations, the Conserver Cult is peculiarly an American phenomenon, probably because no other country provides the affluence, knowledge industries, communications capabilities, and wealth of natural resources necessary for its cultivation. There are no strictly comparable movements in other countries of the West, while Communist countries of the East would not tolerate such a movement for ten minutes. However, although the cult is homegrown, it would be something less than realistic to suppose that the subject of its interest—energy—would escape the notice of the Soviet Union. There is testimony, in fact, that as long ago as the mid-fifties the Soviets were more than a little interested in this subject. Russian physicist and human rights activist Andrei Sakharov, for example, has written of

the statements of an important Soviet official who I heard speak in 1955 when "they" still considered me one of "theirs." It involved the reorientation of Soviet policy in the Near East, the support to be given to Nasser in order to stir up an oil shortage in Western Europe and thus to acquire a more efficient means of pressure. At present, the situation is much more complicated and more nuanced. But the USSR still has a political interest in exploiting the energy difficulties of the West. Is the USSR (or other Eastern countries) currently instigating an anti-nuclear campaign? I have no credible information on this subject. But if it were so, the anti-nuclear prejudices and the incomprehension of the inevitability of the nuclear age are so widespread that a few insignificant and imperceptible efforts would be enough to strongly influence the scope of these campaigns.[22]

Since Sakharov penned these words in 1977, more information has come to light. In July 1979, for example, *Red Line,* published by the Cardinal Mindszenty Foundation in St. Louis told of a "report just published by the Institute for the Study of Conflict" which "reveals that the Communists and other extreme left-wing groups are mounting a world-wide propaganda campaign in the West against the development of nuclear power. The authors of *Nuclear Power: Protest and Violence* stated that 'the Soviet Union derives direct benefit from Western weaknesses, and clearly the anti-nuclear campaigners' tactics are open to exploitation. It already enjoys the strategic advantage of enriching a sizeable proportion of Western Europe uranium. It is the only country apart from France that is developing a Fast Breeder Reactor programme; it brooks no opposition to its nuclear programme, which it hopes to expand rapidly with the help of Japanese technology. If as a result the USSR does not experience an energy gap in the 1980s, the West's present technological lead will be diminished and possibly even disappear.'"[23]

It should therefore come as no surprise that the September 1978 issue of *Political Affairs,* the U.S. Communist Party's theoretical journal, announced that "the movement against nuclear power deserves the active support and participation of the Left, including the Communist Party. The Communist Party has been against the construction of additional nuclear generating facilities in the U.S. so long as these plants would be under the control of private energy monopolies and run for profit."[24] Meanwhile, the Trotskyite Communist weekly, *The Guardian,* applauded *The China Syndrome* as "a propaganda event of the first order," and the U.S. Communist paper, the *Daily World,* stated the anti-nuclear movie was "the most contemporary film ever made. While its message is alerting millions of people, work should begin at once on its sequel, hopefully to channel our emotional responses into effective political action. . . ."[25]

Zoltan Glatter, a Hungarian refugee and executive member of the Central European Federalists, puts all of this into perspective when he writes that "it is not surprising that one finds the extreme left in most Western countries bitterly opposed to nuclear power whilst their friends behind the Iron Curtain are launching a massive construction program. Nuclear power offers the Communists very fertile ground to further subvert and disrupt the economies of the Western world. The classic ingredients exist—hook on to a movement founded originally on genuine concern, infiltrate it and then motivate it for political ends." Strategically, Glatter sums up, "nuclear power is more important to the West than it is to the Soviet Union, yet from present trends one would think the reverse were true. If a healthy pace of nuclear power development is not maintained, the West must increasingly rely on energy imports. The Kremlin knows this and with her naval buildup threatening the supply routes, and her energy independence strategy, of which nuclear power will play a major role, the Soviets will perhaps not need the 'other nuclear weapon' to impose her will upon us."[26]

It is particularly ironic that the Soviet Union would be found propagandizing for "safe" nuclear power in Western countries since, as Petr Beckmann points out, up until "recently (and possibly later), the USSR built its nuclear power plants without emergency core cooling systems and without containment buildings; this may have been based on the indisputable Soviet experience that it is easier to make new Russians than new reactors."[27] However, there is no reason to wonder why responsible Western observers of Soviet Russia report that "few aspects of American activity draw more jubilant reactions behind the Iron Curtain than the successes of the antinuclear movement."[28]

17

GROW OR GO!

The energy crisis confronting the United States today is the result not of insuperable scientific or technical problems but an ideological power grab. This power grab serves a number of constituencies ranging from those who would create a low-energy, low-technology, decentralized utopia of the future to those who would erect a totalitarian police state of the present. But, in either case, the result would be catastrophic for the American people.

However, as the nation enters the eighties, the Conserver Cult is riding higher than ever. Denis Hayes, organizer of Earth Day and Sun Day and formerly of Worldwatch Institute and the Solar Lobby, was appointed director of the Solar Energy Research Institute by the Carter administration. Hayes immediately demanded $100 million for a building and more taxpayers' money than went to the space program to make 25 percent of U.S. energy consumption solar over the next two decades. "Long before World War II," comments Petr Beckmann, "the scientific institutions of the U.S.

government attracted some of the greatest scientists. . . . National research labs have not only employed some of the best scientists, but they have also been directed by men respected by their colleagues for their scientific achievements. That tradition has now been rudely upset. . . ."[1]

S. David Freeman, political architect of the Conserver Cult, is serving as chairman of the Tennessee Valley Authority, the country's largest electric utility and a major operator of nuclear as well as coal-fired powerplants. Previously celebrating Sun Day, Freeman "appeared before a crowd in Memphis wearing bluejeans, a turtle neck sweater and an extremely bright yellow T-shirt over it. He brought applause from the environment-conscious crowd by suggesting that sufficient use of solar energy could eliminate the need for a future nuclear power plant."[2] Freeman made his soft energy strategy even clearer when as TVA chairman he banned from TVA's public displays a number of articles discussing nuclear power, including one by pronuclear former TVA chairman Aubrey ("Red") Wagner. Freeman issued a memo, or "Dave-Ogram," as staff members called it, reading: "I found this pro-nuclear propaganda and policy views of Red Wagner that I disagree with dominating the display outside our energy R&D office in Chattanooga. There was nothing on solar energy. I'm not opposed to nuclear power, but you should know by now that this material does not represent my views. Please take immediate action to bring our publications up to date and stop handing out material you know I can't stomach."[3] Commented the *Knoxville Journal:* "Actually, the bulk of the material he [Freeman] purged represented rather balanced, objective views on the topic of nuclear energy. Their removal thus denies school children and others desiring to learn more about energy a potentially valuable source of information. We find closedmindedness objectionable on the part of any public official. We find any effort to compel others to close their minds—by burning textbooks or other means—unthinkable in a free society. . . . Put 'em back, Dave. . . ."[4]

Meanwhile, Amory Lovins, accused of basing his opposition to nuclear power and his preference for "greatly increased energy productivity and appropriate renewable energy sources" on "impliedly Marxist and manipulative" ideology, responded, in effect: "Who, me?" Stated Lovins, presumably with a straight face, "More careful readers of my work will know that my arguments rest solely and explicitly on classical criteria of economic and engineering efficiency, and that their tacit ideology is the most orthodox market theory. I do suggest that some energy policies are likelier than others to have prohibitive political costs, but these arguments of political economy are independent of and subordinate of my economic ones. Indeed, my view that a soft energy path is a way to *avoid* social change is often uncongenial to the left, who fault me for *not* using energy policy as a vehicle for ulterior motives."[5]

As if to prove Lovins' point, a six-year energy project of the prestigious Harvard Business School resulted in a book called *Energy Future,* which essentially bought the Conserver Cult vision of the future lock, stock, and barrel. Edited by Professor of Business Administration Robert Stobaugh and Kennedy School Lecturer Daniel Yergin, the book was acclaimed by Lovins, who enthused that "from the Harvard Business School comes a lucid, compelling, and authoritative argument that business as usual doesn't work. Instead, the Harvard iconoclasts confirm—with impeccable market economics—that conservation and small solar technologies are the cheapest, quickest, and surest energy sources. This book is an incisive debunking of myths and goes a long way towards solving the energy problem." Jim Harding, energy director for Friends of the Earth, also approved, stating that "we've been hearing for years that there is no energy crisis or that the only solution is more nuclear power plants, more coal, more oil—in short, more of the same. Stobaugh and Yergin are lucid and forceful, buying neither conventional wisdom nor easy remedies. They make a solid dollars-and-cents case for conservation and solar power in a way that will persuade the people who run the country."[6]

Coeditors Stobaugh and Yergin specifically deny that they look on energy as a way to achieve the Conserver Cult vision of a future "soft" society, stating that "those with a stake in conventional wisdom about conventional energy sources may charge the conclusions of this book are unrealistic, that the authors are romantics opposed to economic growth, or that we advocate basic changes in the way the society is organized. We are not and we do not. Indeed, we do not side with those romanticists who have a vision of the national life decentralized in many spheres through the mechanism of the energy crisis to the point where it becomes a post-industrial pastoral society. For we wish to see the system prosper and in a vital way, and this means greater reliance on the free market." But they go on to say with typical Lovinsian rhetoric: "still, we do not subscribe to the views of the other set of even more powerful romanticists—industrial romanticists, who believe that it is possible to return to an era of unlimited production and that production alone can be the nation's salvation."[7] Stobaugh and Yergin also give a deep, albeit discreet, bow to Lovins in their acknowledgements when they say that "late in this process [of writing the book], but fortunately not too late for the book, we had the invaluable experience of working with William Bundy, editor of *Foreign Affairs,* who has been editor of some of the truly significant energy articles of the 1970's."[8]

However, one will look in vain in *Energy Future* for what Lovins calls "impeccable market economics." What one will find instead is what might be called "selective problem-solving." In the hands of Stobaugh and Yergin, this technique consists of throwing up one's hands and claiming that problems confronting production of conventional energy sources such as oil, natural gas, coal and nuclear power are unsolvable, while simultaneously proposing an overflowing grab-bag of solutions to problems facing conservation and solar power. Thus, Stobaugh and Yergin tell us that "there is no domestic oil solution to the problem of increasing U.S. oil imports, no way that production from American oil wells can close the gap of nine million barrels daily between what the United States produces and what it consumes."[9] They tell us that "the nation should not plan on greater quantities of natural gas to stop the rise in oil imports. Indeed, it will be a challenge to find enough new gas reserves to maintain production at current levels."[10] They tell us that "despite its much-touted abundance, coal will not become our major near-term solution to the energy problem," although "its use . . . will grow, and it will play an increasingly important role in certain sectors."[11] They tell us, in italics no less, that *"in any case, nuclear power offers no solution to the problem of America's growing dependence on imported oil for the rest of this century."*[12] And they sum up, "It is clear that domestic oil, gas, coal, and nuclear cannot deliver vastly increased supplies, although it is equally clear that these sources cannot be ignored. America needs all of them. Broadly speaking, however, the nation has only two major alternatives for the rest of this century—to import more oil or to accelerate the development of conservation and solar energy. This is the nature of the choice to be made. Conservation and solar energy, in our view, are much to be preferred."[13] But why prefer conservation and solar energy? Don't they also face problems? Well, yes, but we're going to solve those problems, say Stobaugh and Yergin. According to Yergin, the "obstacles to conservation are manifold." But never fear, because "to overcome them in a politically acceptable and nondisruptive way" will simply require "adroitness." The "movement toward greater energy efficiency, toward greater tapping of conservation energy, will be governed by a complex interaction between government and society. A public policy is required that shapes strong coherent signals, all of which point in the same direction."[14]

As far as solar energy is concerned, Modesto A. Maidique, a Harvard Business School assistant professor who is one of Stobaugh's and Yergins's authors, admits that "there are two barriers, economic and institutional, that must be overcome for solar heating to make a significant contribution." He notes that the solar energy industry, in the words of a Massachusetts state government official, is an "elitist phenomenon." He quotes the marketing manager of a leading solar firm, who described the 1977 solar consumer as "a man in his late forties or fifties, typically an engineer or an architect earning forty to fifty thousand dollars a year. He buys equipment for status or philosophical reasons. Economics are not a factor." He goes on to say that "our surveys support the contention that the rate of return, perceived as too low, constitutes the critical economic obstacle." But he adds that "it need not be."[15]

How will we solve the problems of conservation and solar energy? According to Stobaugh and Yergin, we will have government "give financial incentives to encourage conservation and solar energy." How big will these incentives be? Well, how about an "incentive payment or other form of offsetting subsidy of two-thirds of the cost of implementing conservation and solar energy"? Was that figure *two-thirds?* Yes, say Stobaugh and Yergin, but "we do not, however, recommend this large an incentive, for we acknowledge readily that our calculations are only crude approximations. Furthermore, for the most part we would think somewhat lower incentives would probably be sufficient to encourage a substantial increase in investment in conservation and solar energy." But why provide financial incentives in the first place? "We favor financial incentives to encourage consumers to use conservation and solar energy," state Stobaugh and Yergin, "not because there is anything virtuous about these energy sources, but because they make good economic sense." That's fine, but how can conservation and solar energy make "good economic sense" when they require a subsidy of up to a two-thirds of their cost? Why not provide the same subsidy to conventional energy sources and see what would happen? A subsidy could be furnished to "conventional domestic resources by giving financial incentives to producers of oil, gas, coal and nuclear power," Stobaugh and Yergin admit. But, sorry, they're not going to recommend that because "we have serious doubts that this program would be politically acceptable." Oh, yes, but what's so politically acceptable and where is the money going to come from to subsidize conservation and solar energy up to two-thirds of their cost? "Many oil and gas executives realize that a windfall tax on part of the profits that result from de-regulation of old oil is likely," Stobaugh and Yergin say. "We would propose that the windfall tax be specifically assigned to financing—primarily by tax credits, but also by grants and loans—conservation and solar energy. . . ."[16] Let's get this straight. Money is going to be taken from profits resulting from the sale of oil in the marketplace and used to subsidize conservation and solar energy *which otherwise couldn't make it in the marketplace,* right? Right. Well, doesn't this have more to do with a Lovinsian political preference for conservation and solar energy than his "impeccable market economics" or the Stobaugh-Yergin "greater reliance on the free market"? Answer: Yes. So much for *Energy Future* and the energy project of the Harvard Business School.

In mid-1979, ecologist Barry Commoner and consumerist Ralph Nader were feeling so heady with success that they decided to go for the brass ring, announcing formation of an alternative political party called the Citizens' Party that was to have a candidate for president in the 1980 election. Columnists Jack Germond and Jules Witcover reported that the "Citizens Party was guaranteed some respectful attention by the press because its guiding spirits are among the leading goos-goos of our time, consumer advocate Ralph Nader and ecologist Barry Commoner, who apparently is going to end up being the presidential candidate."[17] The *Wall Street Journal* reported that the "populist" platform of the Citizens' Party "calls for solar power, more citizen control over major corporations and a 'swift halt' to nuclear power. It also wants government-guaranteed jobs, price controls and nationalization of the oil companies."[18] Commoner put it more bluntly when he announced the new party, stating that "the power of the corporations in America today is greater than the power of the citizens of this country . . . and we think it's time to reverse the situation."[19] And, when Commoner says "reverse the situation," he means go all the way to a socialist economy and nationalization of the energy industry, among other objectives. This follows from his book, *The Poverty of Power,* wherein he lays out a scientifically tortured and economically upside-down path to his final conclusion that "the pervasive and seemingly insoluble faults now exhibited by the United States' economic system can best be remedied by reorganizing it along socialist lines. . . . It seems unrealistic . . . to categorically reject a socialist economy on the grounds that its political form is necessarily repressive and therefore abhorrent to the democratic freedoms that are the foundation of political life in the United States. That no existing example of a socialist society—whether the U.S.S.R., China, or Cuba—is consistent with both the economic democracy of socialism and the political democracy inherent in U.S. tradition means that wholly new political forms would need to be created. . . . Only a few years ago *nationalization* was a term that brought averted eyes or nervous titters in Washington. Now, out of necessity, it is regarded—in a very gingerly fashion, it is true—as one possible way to deal with the problems of the railroads and even of the energy industry."[20]

Meanwhile, columnist Jeffrey St. John had previously analyzed the anticorporation attack led by Ralph Nader and embodied in the Citizens' Party platform in such promises as to achieve "more citizen control over major corporations." St. John concluded that the efforts of Nader and his allies to "institutionalize as a matter of public policy a philosophy of no-growth for the private economy are nothing short of intellectual and moral vandalism. It is an attempt to seize by degrees with the power of the state the private mechanism. In short, it is a prescription to return us to the state of savages and serfs."[21]

But, if one, new, antinuclear, anti-free-enterprise, political party is good, two are apparently even better. Thus it was that columnist Paul Scott reported that former UN Ambassador Andrew Young "is privately considering a key role in the launching of a major, new third political party should Carter fail to win the Democratic party nomination. Young's political point man— the Rev. Jesse Jackson, of the Chicago-based PUSH organization—is already testing the waters for the new 'far left' based political action group. Under the banner of the Campaign For An Economic Democracy, Jackson will join some of the most vocal radicals of the 1960s and early 1970s for a whirlwind, political barnstorming tour of the key presidential primary states. . . . Joining Jackson will be Jane Fonda and her husband, former anti-war activist Tom Hayden; William Winpisinger, president of the International Association of Machinists; Cesar Chavez, founder of the United Farm Workers; Eleanor Smeal, of the National Organization for Women; and a dozen national and state 'far left' lawmakers. Their national political tour will begin with a big rally in Harrisburg, Pa." According to Scott, the Campaign For An Economic Democracy has close ties with "the recently organized and 'socialist-inspired' Citizens Party headed by ecologist Barry Commoner and backed by Ralph Nader, the Washington-based 'consumer' advocate. . . . Significantly, the platforms and position papers of the two groups are very similar on major issues. Both groups, for example, advocate the end to the development of nuclear power, cutbacks in military spending, guaranteed jobs for all willing to work, new government controls on corporations, and an expanded policy of accommodating Moscow, Peking, Havana, and Hanoi."[22]

In late 1979, a Big Oil Protest Day was held by the "Campaign to Lower Energy Prices," a coalition of more than 200 labor, consumer and citizen groups, which not only wants to "break up the oil giants but . . . name a special federal prosecutor to investigate them and establish a public energy corporation to explore and develop oil." Organizations supporting the campaign included the International Association of Machinists, the National Education Association, the National Council of Senior Citizens, the United Auto Workers, the United Steelworkers of America, Americans for Democratic Action, Jesse Jackson's Operation PUSH, Ralph Nader, the Democratic Socialist Organization Committee, and the Environmental Policy Center. The coalition passed out cards recommending that the following message be sent to Congress: "I'm tired of oil company rip-offs, windfall profits, and higher prices. I support the Citizens' Energy Program to put a lid on oil and gas prices, stop phony shortages by the oil companies, appoint a special prosecutor, and end the oil monopoly." Said Nader: "We want to reach all those who are angry, but who feel buried by the bigness of the oil industry and bullied by callous politicians who have allowed them to suffer under monopoly energy prices."[23] Thus was the Conserver Cult and its growing number of witting or unwitting allies marshaling every demagogic issue at their command to go for broke in an attempt to gain power over energy and, ultimately, the lives and lifestyles of the American people as the nation entered the decade of the eighties.

However, there are at least several roadblocks in the cult's path to power. First of all, there are the facts of energy. Contrary to cult claims, a growing supply of energy is vital to the continuing socioeconomic advancement and national security of America. The United States has vast domestic energy resources which can be developed to meet future needs. These energy resources can be developed in an environmentally acceptable way with minimum safety risks at rising but, in comparison to the value of the resources to society, still reasonable prices. Rising prices will automatically create financial incentives to not only increase conventional energy production but expand energy conservation practices and develop new, alternative energy resources. Growing domestic energy production can best be achieved at the least possible cost by private energy companies operating in a profit-oriented, competitive marketplace. Financial problems of low-income people resulting from rising energy prices can best be alleviated by programs specifically designed to meet this objective. Following this path to domestic energy production will place ultimate power over energy production, socioeconomic goals and lifestyles of the future in the hands of the American people where it belongs rather than in the grasp of the power-hungry Conserver Cult.

Another roadblock in the Conserver Cult's path to power is the fact that Americans are increasingly getting fed up with its doomsday, need-for-no-growth pronouncements. Louis Harris, for example, reported in late 1979 that "the demands of environmentalists, which have gained increasing public support over the last generation, are now seen as 'making it more difficult to produce adequate energy for the country.' A 57 to 27 percent majority of Americans believes this to be true. There is a widespread sense that energy self-sufficiency in the United States and pursuit of an all-out course to preserve and clean up our air and water, are now in real conflict. It is not that Americans have turned their backs on either the environmentalists or the antipollution groups. Instead, Americans are willing to bend a little on environmental priorities, and they feel that the environmentalists are not as willing to adapt to the new energy necessities." According to Harris, by "49 to 35 percent, a plurality of Americans feels that environmentalists are 'out of touch with the public' and do not 'reflect public feelings.' By 56 to 28 percent, a majority also feels that the leaders and spokesmen of the environmental movement do not 'consider the cost of what they are asking for.'"[24] Harris reported similar majority support for nuclear power development even after the Three Mile Island accident, which resulted in slippage from an even greater level of support in the past. By 68 to 29 percent, a majority would "allow nuclear power plants to be built, if the government regularly inspects the plants to be sure there is no radioactive leakage." By 67 to 29 percent, a majority also would allow "nuclear plants to be built, if the plants met tough government standards for nuclear waste disposal." By 58 to 36 percent, a majority would allow "nuclear power plants to be built, if the government is satisfied on inspection that an accident is unlikely to happen."[25]

This continuing demand of the American people for increased domestic energy production and their growing recognition that energy development is being blocked at the expense of the public interest is now finding a growing political voice in both Congress and state capitals all over the nation. "Growth will be the key issue for the remainder of this century, and it is the resolution of that issue which will determine the lifestyles of most Americans for generations to come," states Senator James McClure. "To fully understand the implications of this debate, it is useful to take a look at who the opponents of

growth are. Tom Wolfe has called them the 'me generation' and Herman Kahn terms them 'the new class.' They
often call themselves consumerists or environmentalists.
Whatever label is used, however, what this group represents is an affluent, politically active, college-educated
minority whose influence is far out of proportion to its
numbers. . . . Economic growth has been inextricably
linked to the growth of the supply of energy throughout
history. The advocates of the stable state economy realize
that if they can control the supply of energy, they can
control the rate at which an economy is to grow. Energy is
the one crucial variable in any economic system . . . if
energy is not available, nothing can make up for its
absence. For this reason, control of the supply of energy is
tantamount to control of the economy. This, in turn,
would give the Federal Government an unprecedented
ability to control individual lifestyles. The concept of
controlling the behavior of the energy-using population,
or for controlling behavior generally, is found throughout
the actions proposed by the new class." However, as
McClure goes on to point out, virtually "every poll taken
regarding energy attitudes has demonstrated that, on
balance, most Americans favor the development of all
energy options, whether they be coal, nuclear, or what
have you. . . . The opponents of the development of our
domestic energy supplies have relied for too long on
rhetoric and emotion to convey their message. They
simply do not have their facts straight. For the most part,
these inaccuracies are so blatant that even the unsophisticated observer can see through them when they are
challenged by accurate data. . . . The American public is
basically optimistic. People want to believe that we can
solve our problems, and are rapidly tiring of the harbin-
gers of doom who seem to see disaster in every action."[26]

Congressman Larry McDonald declares that "there is a new war raging in America today. Following their success in surrendering Southeast Asia, the radicals of the Vietnam War era have embarked on another cause which could well destroy the very fabric of American life. Their target is our energy production capability with nuclear energy their number one objective. By halting nuclear energy, they would roll back the clock of the American Dream to the horse and buggy, candle power and hard labor days. Rather than truth and facts, anti-nuclear zealots promote discredited studies, million-to-one accident scenarios and a host of trumped-up scare tactics in their efforts to permanently close all nuclear plants in America. . . . Energy is the lifeblood of our economic system. Nuclear energy has proven cost effective, practical on a large scale and safe, but it is the focus of the no-growth revolutionaries. All other energy sources will be easy prey after the cleanest and safest is effectively killed. In short, nuclear energy is only the beginning. Their goal is to halt all growth in our nation by destroying our energy production ability and no amount of reasoning or negotiating can change their minds."[27]

Congressman Mike McCormack points out that

early in 1977 President Carter proposed that we limit our total energy growth rate to an average of less than 2% per year during the balance of this century. This is an extremely ambitious and optimistic goal. Our growth rate since 1950 has been between 3 and 3.5% per year. Continuing this long range growth rate would, according to the president, cause us to consume the equivalent of about 84 million barrels of oil/day (from all sources) in the year 2000. The President's goal is to reduce this by one-third: to 56 million barrels of oil/day equivalent consumption in the year 2000. The startling fact of the crisis we face is that if we are successful in meeting the president's optimistic goal, and do so while continuing our present utterly unacceptable rate of petroleum imports, we would still be required to increase our domestic energy production capacity by 70% in the next 21 years. If we assume little or no imports in the year 2000, then meeting the president's goal of 56 million barrels in the year 2000 would require doubling our present domestic energy production capacity, during this 21-year period. . . . So we are facing at least economic and political crises if we try to import the oil we will need, and we are facing domestic economic and political crises if we do not continue our imports unless—and this is the one clear message—unless we dramatically increase our domestic energy production capacity.[28]

Says Governor James A. Rhodes of Ohio:

We have a serious energy problem that is at the root of other major national problems. . . . It is repugnant to Americans to think of themselves as a second-class nation. But more and more, we must face the fact that we no longer are in complete control of our own destiny. . . . The alleged shortage of oil in our nation is a farce, the inevitable result of emotional environmental extremism which effectively limited other sources for generating power on the eastern seaboard. . . . We were forced into this position by professional environmentalists, particularly on the east coast. They stopped offshore drilling. They stopped the burning of coal. They have thrown every roadblock in the path of nuclear energy. And, because of their opposition, not a single new refinery has been built on the east coast in the last 22 years. For the good of our nation and our people, we must restore sanity to our national energy policy. We must use our plentiful native coal everywhere that we can, and save oil for transportation. . . . There is no greater priority facing our nation. The manmade, environmentalist-induced energy crisis is at the heart of our inflation problems, our foreign policy problems, and our employment problems. Until it is corrected, these problems will not go away.[29]

Governor James R. Thompson of Illinois puts it this way:

It's high time that our national leaders in the White House and in Congress *stopped* beating around the bush and *started* beating the odds against us on energy. Two years ago, when we *should* have been making hard choices, we didn't. We were offered a program of generalities, and Congress took two years to pass even less. And today, we're importing more oil than when we started and exporting more dollars and jobs and national pride than ever before. We've *got* to make those hard choices. And we've *got* to do it now. It's going to take far more than a program of conservation that fries us in the summer and freezes us in the winter. We've *got* to look hard at the trade-offs between the ideals of a perfect environment and the realities of an industrial society. We don't want to go back to the days when smog blocked our view across the street. But we do want to make our environmental laws more rational. So it *doesn't* take eight years to get a permit to sink an oil well. So it *doesn't* take 750 pieces of paper to run a pipeline across three states' borders. So it *doesn't* take five years of bureaucratic agony to build a factory or ten years to get a coal-fired power plant on line. We have more unused energy under our soil than many times the energy under OPEC nation sands. We need the will to find our way clear to *use* it. . . . We can only hope that the Congress will now get off its duff and *do* something for a change not only to grapple with energy but to help renew our confidence in our government's ability to cope with any and all of the problems that stand between us and the quality of life we want for ourselves and our children.[30]

And, from the State of Maine, State Treasurer Jerrold B. Speers, says that "there is indeed a crisis of confidence in this nation, but it is not a lack of confidence by the people in themselves or in their future. It is a very real and growing lack of confidence on the part of the people not that they will be able to solve their own problems but that they will even be allowed to try. It is a growing lack of confidence that their own government will allow them to plan and live their own future. It is a growing lack of confidence in the intention and intellectual ability of the government to foster and maintain the economic and political freedoms that citizens have enjoyed in this nation for two centuries. That I believe is the real threat to our democracy."[31]

Some members of the media are also increasingly questioning the goals and objectives of environmentalist, consumerist, "public-interest" and other Conserver Cult groups, while pointing out the need to expand domestic energy production. Journalist Richard Rovere in his "Letter from Washington" column in *The New Yorker,* for example, writes that

> advocates of the small-is-beautiful view make many telling points. What is large is, as often as not, ugly and wasteful. There is no merit in growth for its own sake. But for most people in most societies, growth is the way out of such miseries as hunger, severe heat and cold, disease, illiteracy, and wars undertaken for plunder. Mere growth cannot alleviate suffering, but it can provide the necessary conditions—capital, infrastructure, employment—for a social approach to alleviation. Growth in itself cannot bring abundance, leisure and convenience, but they are seldom to be had without it, and to oppose growth on the ground that it is aesthetically offensive or that we would all be better off leading simpler lives is to take a rather callous view of the human condition in those parts of the world—including sections of this country—where life tends to be simple indeed. To ask the poorer countries to conserve oil and to eschew nuclear energy is to ask them to accept continued poverty as a condition of their existence. To ask Americans to mark time until solar energy comes into their homes and factories is to resign ourself to a rate of unemployment . . . that most find intolerable.[32]

T. R. Reid writes the *Washington Post Magazine* that "the world of public interest lobbying today is indisputably in decline. Partly because of a general climate of skepticism, and partly because of the public interest sector's own excesses, the groups are losing their own image as the good guys in town . . . public interest organizations have to be continually watchful for new trends in liberal thought. Sticking to last year's project can dry up funding sources and that could prove fatal." According to Reid, the "current most-favored subject seems to be opposition to nuclear power, and scores of diverse groups are climbing on that bandwagon." Reid, however, does not predict the end of the "public interest" movement. "Those who are not leaving look more and more like a permanent cadre who will spend their careers moving from one public cause to the next. . . . For such people the means will inevitably become the end. The public interest industry will be important not just for its high-minded objectives, but also as a source of income, self-esteem and security. Such a cadre already exists."[33]

Says Peter Stoler in *Time* magazine:

> In short, after weighing the alternatives, nuclear power is necessary. Why, then, the opposition? . . . The opposition reflects a doubt that growth, once the watchword of the can-do American philosophy, is good. The skeptics ignore the reality that a slow-growth or no-growth philosophy could kill the promise of upward mobility. That may be acceptable to the middle- and upper-income people who dominate the antinuclear movement. But it would condemn the poor and the jobless to a perpetuation of their have-not status and could well endanger the future of American democracy, in which the social and economic inequalities of the free system are made tolerable by the hope of improvement. . . . Irrational opposition to nuclear power can only delay a solution to America's energy problems. . . . Even conservative estimates are that the U.S. will need 390 nukes to provide at least 27% of its electric power by 2000. The time to start building these plants is now. Otherwise, they will not be ready when the nation really needs them.[34]

Author and columnist John Chamberlain comments that "I happen to believe that the environment is important and that we must find ways of sharing the earth with the caribou and the bald eagle, the lion and the kangaroo. But we have been taken for a terrible ride by idiots in the name of such things as environmentalism, ecology, and consumerism. With their anti-technology, anti-business bias, the fanatics have compounded shortages, stopped capital investment, and even made it impossible for the nations of the West to follow a sensible foreign policy vis-a-vis Communists and Arabs."[35] Columnist Nick Thimmesch is even more to the point, stating that "Americans must make up their minds. If we are willing to consume less, so be it. If we want more domestic energy, the super-environmentalists and anti-nuclear crowd must be defeated."[36]

Major organizations in American society also are joining the chorus of cries for greater domestic energy production. Here, for example, is how Jacob Clayman, secretary-treasurer of the Industrial Union Department, AFL-CIO, puts it: "We must explore every avenue of actual and potential energy supply. None must be overlooked. That includes coal, solar, hydroelectric, geothermal and NUCLEAR. What of nuclear? Is it safe? One fact impresses me. We have 72 nuclear generating installations in this country. There are 200 more in the free world. That amounts to 30 years of experience. But the fact is that we do not have one known death from nuclear radiation. That is a very important fact. Our economy, our job structure, our way of life rest on an adequate supply of energy. We must forge ahead with nuclear. We must prove to the public that nuclear is safe. We must cut down on the time required to put a nuclear plant in operation— from the present 11-12 years to 6-7 years. This would save $200-$300 million per installation. It's time to get going."[37]

Allan Grant, president of the American Farm Bureau Federation, declares that

the immediate need remains for the development and implementation of a comprehensive national energy policy. This policy should, first of all, define the energy needs of this nation in an accurate and unbiased fashion, something not heretofore done. Next, this policy should examine how these needs can best be met, consistent with the wise use of our natural resources. Absolute top priority within this policy must be recognition that the market system is the only effective means for locating, developing and allocating energy supplies. To these basics, farm and ranch people would add that environmental standards must be realistic and practical. We would call for economic impact statements to force federal and state regulatory agencies to go through some of the same harassment now reserved for others. A sound national energy policy must be realistic about additional energy sources. Remember, farmers are the original users of solar energy. Solar energy powers all of our crops and "manufactures" what we feed our cattle. We view the development of solar, tidal, geothermal, and waste conversion projects with great interest. But we are a practical people. We know this country can't wait another 25 or 50 years for these fledging projects to come "on line." Electricity is here and now. There are numerous sources of energy for the manufacture of electricity. And these resources are far from being developed to their full potential. Atomic energy is here and now. Petroleum and natural gas are here and now. Coal is here and now. If we are to be realistic, the continued expansion of nuclear generating plants, including the controversial breeder-reactors, must be encouraged. Instead, nuclear development has been so greatly discouraged that as many as 50 proposed nuclear plant sites have been abandoned in recent years. To continue to forestall nuclear development is to invite serious trouble. It is asinine to deliberately drop behind in a technological development which holds the key to our brightest energy future. It is not to our credit that we have allowed nuclear energy—the cleanest, safest, almost inexhaustible [energy source]—to become the terror object of the peace-cult and the subject of an environmental, social and political witch-hunt. Ignored in the scare talk about nuclear waste disposal is the fact that many of the so-called waste products are themselves valuable resources and, in every case, they can be housed with no hazard to the environment. But a competitive business environment must be at the heart of any realistic national energy policy. Prices of energy supplies, with a few possible exceptions such as national emergencies, must be allowed to move freely in response to supply and demand and to profit and loss. Nothing less than full deregulation of natural gas and crude oil prices will allow this country to develop and maintain adequate energy supplies and get them distributed on an equitable basis.[38]

And the National Energy Conference of the National Association for the Advancement of Colored People reported that

we note historic direct correlation between the level of economic activity and energy availability and consumption. Energy supply development throughout our nation's history has been critically important to economic growth. We find it very disturbing to contemplate a future in which energy supply would become a constraint upon our ability to solve these critically important social and economic problems which confront Black Citizens. . . . We think there must be a more vigorous approach to supply expansion and to the development of new supply technologies so that energy itself will not become a long-term constraint, but instead can continue to expedite economic growth and development in the future. All alternative energy sources should be developed and utilized. Nuclear power, including the breeder, must be vigorously pursued because it will be an essential part of the total fuel mix necessary to sustain an expanding economy. Other alternative sources, such as solar, geothermal, biomass, tidal, oil shale and synthetic fuels from coal must also be developed and made commercially available at the earliest possible time. A more positive attitude by the Administration toward supply development is extremely important because future developments will be largely determined by the Policy choices being made now by the Administration and the Congress.[39]

343

Mrs. Margaret Bush Wilson, chairman of the NAACP, also spoke out sharply against politicizing the energy situation, stating that "we do not want political decisions when we need sound, realistic ones to deal with problems. A person whose opinions I respect said to me recently that it is the senator from Massachusetts who is controlling energy policy these days. This comment made little sense to me until it was explained that apparently this Administration at the moment is practically immobilized on this subject of energy policy, lest, with an eye on the elections of 1980, it triggers action which may cause Senator Edward Kennedy to sound off with a round of public criticism. If this is true, I say to this Administration, and to Senator Kennedy: A pox on both your houses."[40]

New types of public interest law centers have also sprung into action in recent years in response in part to attacks on domestic energy production. One such center is the Washington, D.C., based National Legal Center for the Public Interest which has regional affiliates in not only the capital but Atlanta, Chicago, Kansas City, Denver, Springfield, Massachusetts, and Harrisburg, Pennsylvania. Others include the New England Legal Foundation in Boston and the Pacific Legal Foundation in Sacramento, California. "We're going to look the Naderites . . . in the eye and we're going to tell them we've been pushed around enough, and now we're going to start fighting back," vows one of the new public interest law centers, the Washington Legal Foundation, whose executive director Daniel Pepeo adds that "I think people are fed up with Ralph Nader. People like Nader and Jane Fonda lead a very rarefied existence. These people have no sense of reality." Representing those who believe among other things that nuclear power is the ultimate answer to the nation's energy problems, the law centers are "out to tame environmental extremists and rabid regulators, and to bring balance to a regulatory process that has tipped in favor of environmental and consumer activists."[41]

James G. Watt, former vice chairman of the Federal Power Commission and now president of the Mountain States Legal Foundation in Denver, expands on the reasons for the formation of the new public interest law centers, explaining that

> because Congress, because our Executive branch of government, because our state legislatures have been so fickle, leaders of various interest groups in America have turned to the courts for dependable even if not desirable direction. The legislation which is passed with the glorious titles is soon forgotten on Capitol Hill and then the bureaucracy does its work. The result is usually conflict forcing the affected people to go to the courts for the final decision. Although the courts are the most dependable segment of our government, some will agree that the courts should not be so deeply involved in our lives. . . . Why are the judges making these sweeping decisions that affect every angle of our lives? Because, for one reason, activist groups have recognized that the major decisions in America are being made in the courts. With millions of dollars made available from foundations, they have hired bright people and have moved aggressively into court to bring a halt to economic development and growth of this Nation. There are many reasons for their startling success, but in many instances these extreme groups prevailed because their "authority" to speak for the "public interest" was not challenged. In the last few years, community leaders have put together a new type of public-interest law center dedicated to individual freedoms, the right to private property and the private enterprise system. These public-interest law centers are demanding that issues dealing with the real public interest are fairly presented to the courts. They are challenging the right of single-purpose groups to claim they represent the public interest.[42]

But, most important, there is today in America a

rapidly growing grass roots movement dedicated to achieving increased domestic energy production. This movement is a living example of the quote from Edmund Burke beginning this book, which reads in full:

> How often has public calamity been arrested on the very brink of ruin, by the seasonable energy of a *single man?* Have we no such man amongst us? I am as sure as I am of my being, that *one* vigorous mind without office, without situation, without public functions of any kind (at a time when such a thing is felt, as I am sure it is), I say, *one* such *man,* confiding in the aid of God, and full of just reliance in his own fortitude, vigor, enterprise, and perseverance, would first draw to him some few like himself, and then that multitudes, hardly thought to be in existence, would appear and troop about him.[43]

Fortunately, there are many such men *and women,* and they have drawn together a growing multitude of people in a diverse number of energy advocacy organizations throughout the country. These groups include Citizens for Energy and Freedom, which attracted more than 15,000 people to the federal government's Rocky Flats plant northwest of Denver, Colorado, for the largest pronuclear rally ever held in the United States, Americans for More Power Sources, Inc., Citizens for Total Energy, Americans for Nuclear Energy, Massachusetts Voice of Energy, New Hampshire Voice of Energy and many others (a more complete list of energy advocacy organizations is provided in Appendix A). Energy advocacy groups came together for the first time in early 1979 in Washington, D.C., at a National Conference on Energy Advocacy sponsored by the Heritage Foundation and thirty-five other proenergy organizations whose common link was a concern about the future availability of energy and jobs in the United States. Anna L. West of the Energy

Research Group relates that more than 700 people attended this "first-ever National Conference on Energy Advocacy. To give a feeling of size, Ralph Nader's 'Critical Mass' meeting, an approximate, anti-nuclear equivalent, attracted less than 600 people. . . . People from labor, business and academia, citizens, housewives and engineers, all came together for a common cause: development of domestic energy resources." Based on this objective, the conference resolved that federal, state and local legislatures and administrations: (1) promptly implement energy policies permitting the increased reliance on coal and nuclear power to provide needed electrical energy reliably and in a timely manner and encouraging the development of domestic oil and natural gas resources; (2) plan for supplying future energy needs by aggressively pursuing research and development for emerging energy technologies; (3) give full weight to views on energy expressed by the large number of citizens who support continued energy growth as an essential part of American energy policy; and (4) critically review the necessity and the structure of all existing and proposed regulations and promptly improve the quality of regulation, while decreasing its cost by eliminating redundancies, inefficiencies, inconsistencies and unwarranted delays in the regulatory process. The conference also resolved that the Congress and federal administration demand that the designated government agencies immediately implement waste disposal solutions and thus remove unwarranted political constraints on the utilization of nuclear power.[44]

The conference basically supported the development of all energy resources—conventional and unconventional, hard and soft—because this is no time to be picky. Each energy source is not only needed, but has a place in future

347

domestic energy production and consumption. As Margaret Maxey puts it, if "our future as a nation is to be an ethically responsible and reliable source of betterment for humankind on this globe, then we must recognize that the bridge to a just, sustainable, environmentally protected society lies in our technological ability to sustain an orderly continuity of organic development of our human and natural resources. We should long ago have abandoned the polarized rhetoric of growth vs. no-growth, hard path vs. soft path, solar energy vs. nuclear energy. We need all the options we can muster in the next fifty years if we are going to avoid socially disruptive and avoidable energy shortages. Technological development has finally made it possible to provide new levels of life-sustaining material well being to everyone on this planet."[45]

Meanwhile, if members of the Conserver Cult wish instead to pursue their utopian vision of low-energy, low-technology decentralization, that of course is their prerogative. The United States is a big, diverse and free country, and one of the incomparable advantages of such size, diversity and freedom is the ability to accommodate widely varying lifestyles. However, it is one thing to choose a lifestyle and live it. It is another thing to impose a lifestyle on someone else who neither chose nor desires it. Stephen J. Tonsor, professor of European intellectual history at the University of Michigan, relates that

> some time before ecology became fashionable and antinomian irrationality became the vogue among the well-healed cognoscenti in our society, C. P. Snow made this point very ably. He wrote: "For, of course, one truth is straightforward, industrialization is the only hope of the poor. I use the word 'hope' in a crude and prosaic sense. I have not much use for the moral sensibility of anyone who is too refined to use it so. It is all very

348

well for us, sitting pretty, to think that material standards of living don't matter all that much. It is all very well for one, as a personal choice to reject industrialization—do a modern Walden, if you like, and if you go without much food, see most of your children die in infancy, despise the comforts of literacy, accept twenty years off your own life, then I respect you for the strength of your aesthetic revulsion. But I don't respect you in the slightest if, even passively, you try to impose the same choice on others who are not free to choose. In fact we know what their choice would be. For, with singular unanimity, in any country where they have had the chance, the poor have walked off the land into the factories as fast as the factories would take them.[46]

Oil . . . natural gas . . . coal . . . nuclear energy, including fission reactors, breeder reactors and fusion . . . new, more efficient and environmentally acceptable methods of burning coal now in engineering demonstration stages such as fluidized-bed combustion and magnetohydrodynamics . . . hydropower . . . geothermal energy . . . solar energy . . . wind power . . . biomass . . . hydrogen which burns cleanly in cars and in other energy applications leaving only water as a residue . . . fuel cells which produce electricity more efficiently by chemical means: All of these can and must be used to insure a growing supply of energy in the future. Nor does this exhaust the list of possible future energy sources. For recent discoveries have been made confirming the existence of a previously unknown subatomic particle called the "gluon," which could open the door to the development of what is being called the "ultimate energy source." *Chicago Tribune* science editor Ronald Kotulak, for example, reports that "the discoveries, which have electrified the world of high-energy physics, may eventually lead to ways of converting matter into pure energy. Such an accomplishment would outshine the sun or any other type of

349

energy and put mankind on an energy Easy Street. The ability to convert the matter in one drop of water into total energy would make energy so cheap and abundant that it would cease to be a consideration in the evolution of the civilized world." Reporting on an interview with Dr. Leon Lederman, director of the Fermi National Accelerator Laboratory, Kotulak continues that "incredibly, it would take only 81 gallons of water to meet all of this nation's energy needs for one year. Furthermore, the 'ultimate energy source' would be safe and clean with no dangerous or unwanted byproducts. Kotulak sums up by quoting Lederman to the effect that "we are talking about the possibility of an energy supply that is fantastically more prolific than anything we have now. Although still a long way off, but perhaps as soon as 30 years, the ability to turn matter into usable energy is not an impossibility. I think ways can be found to do it. With such titanic powers, it would be possible to convert anything—rocks, trees, dirt, or garbage—into a limitless new energy source."[47]

However, time—the only truly unrenewable resource—is running out on domestic energy production in the United States. What is critically needed is strong, decisive political leadership to activate the great and growing base of support among the American people for energy development. This is absolutely vital to not only continue socioeconomic advancement but maintain and enhance national security. General Alexander Haig, former supreme allied commander of NATO forces in Europe, for example, warns that "the Russians' success at establishing solid footholds in a vast area—from Afghanistan through South Yemen to both coastal areas of Africa—could at any given moment put in jeopardy the lifelines of Western access to raw materials."[48] And former Air Force Secretary Thomas C. Reed warns that "within five years—a period of inevitable energy confrontation—the Soviet Union is headed for superiority in every measure of nuclear arms."[49]

The United States must grow in domestic energy production or go down the drain of history for lack of energy capabilities. Andrei Sakharov best puts the issue in perspective:

> People must have the possibility, but also the knowledge and the right, to balance the interrelated economic, political and ecological problems against each other soberly and responsibly. Problems concerning the development of nuclear energy and the alternatives of economic development must be solved without baseless emotions and prejudices. It is not merely a question of comfort, not merely of the maintenance of the so-called "quality of life." The issue is far more important—economic and political independence, preservation of freedom for our children and grandchildren. I am convinced that the correct solution will ultimately be reached.[50]

Appendix A

ENERGY ADVOCATES

This appendix provides a representative list of energy advocacy organizations compiled from information furnished by the Heritage Foundation, the Atomic Industrial Forum, and Charles Yulish of GREAT (Grass Roots Energy Alliance Team). These organizations have been formed by individual citizens, energy industry employees acting as private citizens, students, teachers, minorities, labor unions, scientists, engineers, and companies and associations in business and industry. Also provided are lists of proenergy public interest law centers, energy-industry professional and trade associations and research organizations.

Energy Advocacy Organizations

Action Committee for Energy
P.O. Box 5848
Washington, D.C. 20014

AFL/CIO Energy Policy
  Committee
815 Sixteenth Street, N.W.
Washington, D.C.

American Association of Blacks
  in Energy (AABE)
1429 Larimar Square
Denver, Colorado 80202

Americans for Energy
  Independence (AFEI)
1629 K Street, N.W.
Suite 1201
Washington, D.C. 20006

Americans for More Power
Sources (AMPS)
P.O. Box 501
Manchester, N.H. 03105

Americans for Nuclear Energy
P.O. Box 23834
L'Enfant Plaza Station
Washington, D.C. 20024

Americans for Rational Energy
Alternatives (AREA)
P.O. Box 11802
Albuquerque, N.M. 87112

Arizonians for Jobs and Energy
Suite 671
Security Center
222 N. Central Avenue
Phoenix, Arizona 85004

California Council for
Environmental and
Economic Balance
215 Market Street
Suite 930
San Francisco, Calif. 94105

California Energy Council
P.O. Box 74
Carmichael, Calif. 95608

Citizen Advocates for Rational
Energy (CARE)
4459 Kilmer Drive
Murrysville, Pa. 15668

Citizens for Energy and Freedom
941 E. 17th Avenue
Suite 3
Denver, Colorado 80218

Citizens for Energy & Jobs
606 Country Lane
Burlington, Wash. 98233

Citizens for Total Energy
5410 Fairway Drive
San Jose, Calif. 95127

Committee to Separate America
from Foreign Energy (SAFE)
Stony Lonesome
Rockport, Maine 04856

Committee to Supply Adequate
Future Energy (SAFE)
P.O. Box 631
Wading River, N.Y. 11762

Concerned Citizens for the
Nuclear Breeder
Box 208
Ruffs Dale, Pa. 15679

Consumer Alert
1091 Post Road
Darien, Conn. 06820
or
P.O. Box 981
Stamford, Conn. 06904

Coordinating Committee on
Energy
Sixth Floor
345 E. 47th Street
New York, N.Y. 10017

Council on Energy Independence
P.O. Box J
Chicago, Ill. 60690

Energy Advocates
Suite 400
Enterprise Building
Tulsa, Oklahoma 74103

Energy Research Institute
First National Bank Building
Suite 670
Columbia, S.C.

Forum for the Advancement of
Students of Science and
Technology (FASST)
2030 M Street, N.W.
Suite 402
Washington, D.C. 20036

Kern Energy Education Program
(KEEP)
P.O. Box 9295
Bakersfield, Calif. 93309

Labor and Industry for Energy
and Jobs Committee
(Life/Jobs)
130 South Third Street
Suite L-6
Harrisburg, Pa. 17101

Leadership Foundation, Inc.
4808 Cleveland Park Station
Washington, D.C. 20008

Massachusetts Voice of Energy
1032 Statler Office Building
Boston, Mass. 02116

Michigan Committee for Jobs
and Energy, Inc.
Suite 309
419 S. Washington Ave.
Lansing, Mich. 48933

Montana Energy Education
Council
P.O. Box 2693
Great Falls, Montana

National Association of
Pro-America
RR#5, Box 419
Monticello, Indiana 47960

National Council for
Environmental Balance, Inc.
(NCEB)
P.O. Box 7732
Louisville, Ky. 40207

National Energy Resources
Organization (NERO)
1901 No. Moore St. (#805)
Arlington, Va. 22209

National Environmental
Development Association
(NEDA)
National Press Building
Washington, D.C. 20045

New Hampshire Voice of Energy
1153 Cilley Road
Manchester, N.H. 03103

New York State Committee for
Jobs and Energy
Independence
211 East 43rd Street
15th Floor
New York, N.Y. 10017

Nuclear Energy for
Environmental Development
(NEED)
P.O. Box 881
San Luis Obispo, Calif. 93406

Nuclear Energy Women
1513 East Falkland Lane
Silver Spring, Md. 20910

Nuclear Legislative Advisory
   Service
P.O. Box 354
Murrysville, Pa. 15668

Oregon Citizens for Economic
   and Environmental Balance
725 Cottonwood Street
Arlington, Oregon 97812

Pennsylvania Society of
   Professional Engineers'
   Energy Action Committee
2806 McCully Road
Allison Park, Pa. 15101

People for Energy Policy (PEP)
P.O. Box 1420
Los Gatos, Calif. 95030

Realistic Alternatives for
   Vermont Energy (RAVE)
RFD 2
St. Johnsbury, Vermont 05819

S.A.F.E.
Box 183A
Route 1
Grandview, Tenn. 37337

Safe, Adequate Future Energy
   (SAFE)
731 Wisconsin River Drive
Port Edwards, Wisconsin 54469

Scientists & Engineers for Secure
   Energy (SE_2)
410 Riverside Drive
Apt. 82-A
New York, N.Y. 10025

Scientists for Enlightenment on
   Nuclear Sources of Energy
   (SENSE)
Brookhaven National Labs
Building 535
Long Island, N.Y. 11973

Secure America's Future Energy
   (S.A.F.E.)
P.O. Box 353
Monroeville, Pa. 15146

Society for the Advancement of
   Fission Energy (SAFE)
P.O. Box 353
Monroeville, Pa. 15146

Society for Environmental,
   Economic Development
   (SEED)
Suite 1519
Inn of Trenton
240 West State Street
Trenton, N.J. 08608

Vermont Voice of Energy
P.O. Box 338
Shelburne, Vt. 05482

Wisconsin Energy Coalition
6227 W. Greenfield Ave.
West Allis, Wisconsin 53214

Women & Energy (WE)
2744 S.W. Sherwood Drive
Portland, Oregon 97201

Women's Energy Forum
5 Adams Road
Bloomfield, Conn. 06002

355

Proenergy Public Interest Legal Centers

National Legal Center for the Public Interest
1101 Seventeenth Street, N.W.
Suite 810
Washington, D.C. 20005

Regional Affiliates:

Capitol Legal Foundation
1101 17th Street, N.W.
Suite 810
Washington, D.C. 20005

Southeastern Legal Foundation
1800 Century Boulevard
Suite 950
Atlanta, Ga. 30345

The Mid-America Legal Foundation
20 N. Wacker Drive
Suite 2245
Chicago, Ill. 60606

The Great Plains Legal Foundation
Board of Trade Building
127 W. Tenth Street
Kansas City, Mo. 64105

Mountain States Legal Foundation
1845 Sherman Street
Suite 675
Denver, Colorado 80203

Northeastern Legal Foundation
Valley Bank Towers
1500 Main Street
Suite 2402
Springfield, Mass. 01115

Mid-Atlantic Legal Foundation
Chamber of Commerce Building
222 North Third Street
Harrisburg, Pa. 17101

New England Legal Foundation
110 Plymouth Street
Boston, Mass. 02108

Pacific Legal Foundation
455 Capitol Mall
Suite 465
Sacramento, Calif. 95814
or
1990 M Street, N.W.
Suite 550
Washington, D.C. 20036

Energy-Industry Professional and Trade Associations and Research Organizations

American Gas Association
1515 Wilson Boulevard
Arlington, Virginia 22209

American Nuclear Energy Council
1750 K Street, N.W.
Suite 300
Washington, D.C. 20006

American Nuclear Society
555 North Kensington
La Grange Park, Ill. 60525

American Petroleum Institute
2101 L Street, N.W.
Washington, D.C. 20037

American Public Power
  Association
2600 Virginia Avenue, N.W.
Washington, D.C. 20037

Atomic Industrial Forum
7101 Wisconsin Ave.
Bethesda, Md. 20014

Edison Electric Institute
90 Park Avenue
New York, N.Y. 10016

Electrical Power Research
  Institute
3412 Hillview Avenue
P.O. Box 10412
Palo Alto, Calif. 94303
  or
1750 New York Avenue, N.W.
Washington, D.C. 20006

Geothermal Resources Council
P.O. Box 1033
Davis, Calif. 95616

Institute of Gas Technology
3424 S. State St.
Chicago, Ill. 60616

International Solar Energy
  Society
American Section
300 State Road 401
Cape Canaveral, Fla. 32920

National Association of Electric
  Companies
1140 Connecticut Avenue, N.W.
Washington, D.C. 20036

National Coal Association
1130 Seventeenth St., N.W.
Washington, D.C. 20036

Solar Energy Industries
  Association, Inc.
1001 Connecticut Ave., N.W.
Suite 632
Washington, D.C. 20036

Wind Energy Society of America
1700 E. Walnut
Pasadena, Calif. 91106

Appendix B

THE CONSERVER CULT

This appendix provides a representative list of environmentalist, conservationist, consumerist and antinuclear organizations.

Center for Science in the Public Interest
1757 S St., N.W.
Washington, D.C. 20004

Citizen's Energy Project
1413 K Street, N.W.
8th Floor
Washington, D.C. 20005

Committee for Nuclear Responsibility, Inc.
P.O. Box 11207
San Francisco, Calif. 94101

Conservation Foundation
1717 Massachusetts Ave., N.W.
Washington, D.C. 20036

Consumer Federation of America
1012 14th St., N.W.
Washington, D.C. 20005

Critical Mass
P.O. Box 1538
Washington, D.C. 20013

Defenders of Wildlife
1244 Nineteenth Street, N.W.
Washington, D.C. 20036

Environmental Defense Fund
475 Park Avenue, South
New York, N.Y. 10016

Environmental Action
Foundation
724 Dupont Circle Bldg.
Washington, D.C. 20036

Environmental Action, Inc.
1346 Connecticut Ave., N.W.
Room 731
Washington, D.C. 20036

Environmentalists for Full
Employment
1101 Vermont Ave., N.W.
Washington, D.C. 20005

Environmental Policy Center
317 Pennsylvania Ave., S.E.
Washington, D.C. 20003

Friends of the Earth
124 Spear Street
San Francisco, Calif. 94105

Izaak Walton League of
America, Inc.
1800 N. Kent St.
Suite 806
Arlington, Va. 22209

League of Conservation Voters
317 Pennsylvania Ave., S.E.
Washington, D.C. 20003

National Audubon Society
950 Third Avenue
New York, N.Y. 10022

National Parks & Conservation
Association
1701 Eighteenth Street, N.W.
Washington, D.C. 20009

National Wildlife Federation
1412 16th Street, N.W.
Washington, D.C.20036

Natural Resources Defense
Council, Inc.
122 East 42nd Street
New York, N.Y. 10017

The Nature Conservancy
1800 N. Kent Street
Suite 800
Arlington, Va. 22209

The Population Institute
100 Maryland Avenue, N.E.
Washington, D.C. 20002

Public Interest Research Group
P.O. Box 19312
Washington, D.C. 20036

Sierra Club
530 Bush Street
San Francisco, Calif. 94108

Solar Lobby
1028 Connecticut Avenue, N.W.
Washington, D.C. 20036

Union of Concerned Scientists
1208 Massachusetts Avenue
Cambridge, Mass. 02138

The Wilderness Society
1901 Pennsylvania
Avenue, N.W.
Washington, D.C. 20006

Worldwatch Institute
1776 Massachusetts Ave., N.W.
Washington, D.C.

Zero Population Growth
1346 Connecticut Avenue, N.W.
Washington, D.C. 20036

Appendix C

ADDITIONAL SOURCES OF ENERGY INFORMATION

This appendix lists private and governmentally operated organizations which conduct and publish research on energy and U.S. government agencies which provide energy information.

Research Organizations

American Association for the
Advancement of Science
1515 Massachusetts Ave., N.W.
Washington, D.C. 20005

American Enterprise Institute for
Public Policy Research
1150 Seventeenth St., N.W.
Washington, D.C. 20036

The Brookings Institution
1775 Massachusetts Ave., N.W.
Washington, D.C. 20036

The Heritage Foundation
513 C Street, N.E.
Washington, D.C. 20002

Hudson Institute
Qualer Ridge Rd.
Croton-on-Hudson, N.Y. 10520

National Academy of Sciences
2101 Constitution Ave., N.W.
Washington, D.C. 20418

National Science Foundation
1800 G St., N.W.
Washington, D.C. 20550

Organization for Economic
  Cooperation and
  Development
OECD Publication Center
Suite 1207
1750 Pennsylvania Ave., N.W.
Washington, D.C. 20006

The Rand Corporation
1700 Main Street
Santa Monica, California 90406

Resources for the Future
1755 Massachusetts Ave., N.W.
Washington, D.C. 20036

United Nations
Publication Sales Section
Room A-3315
New York, N.Y. 10017
or
Unipub
Box 433
Murray Hill Station
New York, N.Y. 10016

U.S. Government Agencies

Consumer Information
Public Documents Distribution
  Center
Pueblo, Colorado 81009

Council on Environmental
  Quality
Executive Office of the President
722 Jackson Place, N.W.
Washington, D.C. 20006

Environmental Protection
  Agency
Public Information Center
Washington, D.C. 20460

Energy Research and
  Development Administration
  (ERDA)
Public Affairs Office
Washington, D.C. 20545

Federal Energy Administration
12th and Pennsylvania
  Ave., N.W.
Washington, D.C. 20461

Federal Power Commission
Office of Public Information
825 N. Capitol St.
Washington, D.C. 20426

Geological Survey (U.S.D.I.)
National Center
Reston, Va. 22092

Health, Education and Welfare
  Dept.
Office of Consumer Affairs
330 Independence Ave., S.W.
Washington, D.C. 20201

National Energy Information
  Center
Federal Energy Administration
1408 Federal Bldg.
Washington, D.C. 20461

National Petroleum Council
1625 K St., N.W.
Suite 601
Washington, D.C. 20006

National Solar Heating &
  Cooling Information Center
P.O. Box 1607
Rockville, Md. 20850

National Technical Information
  Service (NTIS)
425 13th St., N.W.
Washington, D.C. 20004

Nuclear Regulatory Commission
Public Affairs Dept.
Bethesda, Md. 20014

Nuclear Safety Information
  Center
Nuclear Regulatory Commission
P.O. Box Y
Oak Ridge, Tenn. 37830

U.S. Government Printing Office
Washington, D.C. 20042

NOTES

Chapter 1

1. Dr. John J. McKetta, *The Energy Crisis Deepens* (Louisville, Ky: National Council for Enviromental Balance, n.d.).

2. "Utilities Hit Trends, Warn of Blackouts," *Chicago Tribune,* September 9, 1976.

3. "Blackouts, Curbs on Power Use Likely by 1979, Utilities Predict," *Miami Herald,* September 18, 1977.

4. Robert Young, "Power Industry Warns of Imminent Shortages," *Chicago Tribune,* September 12, 1978.

5. *Eighth Annual Review of Overall Reliability and Adequacy of the North American Bulk Power System* (Princeton, N.J.: National Electric Reliability Council, 1978), p. 11.

6. "Two New Reactors Ordered by Commonwealth Edison," *Nuclear Info,* January 1979, Atomic Industrial Forum, Inc., Washington, D.C.

7. Louis Harris, "An Energy Plan Sought," *Chicago Tribune,* June 1, 1978.

8. Louis Harris, "A Call for More Oil Production," *Chicago Tribune,* January 12, 1979.

9. Louis Harris, "An Energy Plan Sought."

10. John J. McKetta, *The U.S. Energy Problem—America's Achilles Heel* (Louisville, Ky.: National Council for Environmental Balance, 1978), p. 13.

Chapter 2

1. Quoted in John C. Whitaker, *Striking a Balance* (Washington, D.C.: American Enterprise Institute, 1976), p. 50.

2. Ibid., p. 2.

3. Ibid., p. 41.

4. McKetta, *The Energy Crisis Deepens.*

5. Quoted in Edward J. Mitchell, *U.S. Energy Policy: A Primer* (Washington, D.C.: American Enterprises Institute for Public Policy Research, 1974), pp. 1-2.

6. Ibid., p. 14.

7. Merril Eisenbud, *Environment, Technology, and Health: Human Ecology in Historical Perspective* (New York: New York University Press, 1978), pp. 61-62.

8. Ibid., p. 68.

9. John C. Whitaker, *Striking a Balance,* p. 95.

10. Eisenbud, *Environment, Technology, and Health,* p. 321.

11. Mitchell, *U.S. Energy Policy,* p. 71.

Chapter 3

1. E. F. Schumacher, *Small is Beautiful, Economics As If People Mattered* (New York: Harper & Row, 1973), p. 14.

2. Ibid., pp. 14-15.

3. Ibid., p. 145.

4. Ibid., pp. 30-31.

5. Ibid., pp. 33-34.

6. Ibid., pp. 34-35.

7. Ibid., pp. 153-54.

8. Ibid., p. 55.

9. Ibid.

10. Ibid., p. 33.

11. Ibid., p. 53.

12. Ibid., pp. 57-58.

13. Ibid., p. 159.

14. Amory B. Lovins, "Energy Strategy: The Road Not Taken?" *Foreign Affairs,* October 1976, p. 65.

15. Ibid., p. 65.

16. Ibid., p. 88.

17. Ibid., p. 66.

18. Ibid., pp. 68-69.

19. Ibid., p. 93.

20. Ibid., p. 66.

21. Ibid., p. 72.

22. Ibid., p. 76.

23. Ibid., p. 77.

24. Ibid., pp. 81–84.

25. Ibid., pp. 84–85.

26. Ibid., pp. 77–78.

27. Ibid., p. 79.

28. Ibid., pp. 91, 92.

29. Ibid., pp. 88–89.

30. Ibid., p. 94.

31. *Progress As If Survival Mattered* (San Francisco: Friends of the Earth, 1977), p. 7.

32. Lovins, "Energy Strategy," p. 95.

33. Ibid., pp. 65–66, 96.

34. *Progress As If Survival Mattered*, p. 7.

Chapter 4

1. "The Plowboy Interview: Dave Brower," *The Mother Earth News*, May 1973.

2. Joe Klein, "Ralph Nader—The Man in the Class Action Suit," *Rolling Stone*, November 20, 1975.

3. Milton R. Copulos, *Confrontation at Seabrook* (Washington, D.C.: Heritage Foundation, 1978), pp. 38–39.

4. Amory B. Lovins, *Soft Energy Paths: Toward a Durable Peace* (Cambridge, Mass.: Ballinger, 1977), p. 71.

5. Lewis Mumford, "Enough Energy for Life," Committee for Nuclear Responsibility, Inc., P.O. Box 332, Yachats, Oregon.

6. Herman E. Daly, "The Steady-State Economy: Toward a Political Economy of Biophysical Equilibrium and Moral Growth," in Herman E. Daly, ed., *Toward a Steady-State Economy* (San Francisco: W.H. Freeman, 1973), p. 149.

7. John N. Cole, "The Future Has Arrived," *The Mother Earth News*, November 1973, p. 83.

8. Lewis H. Lapham, "The Energy Debacle," *Harper's*, August 1977, p. 61.

9. S. David Freeman, *Energy: The New Era* (New York: Vintage, 1974), p. 8.

10. Ibid., p. 6.

11. Ibid., p. 4.

12. Ibid., pp. 6–7.

13. Ibid., p. 7.

14. Ibid., p. 9.

15. Ibid., pp. 80–81.

16. Ibid., p. 303.

17. Ibid., p. 79.

18. Ibid., p. 84.

19. Ibid., p. 81.

20. Ibid., pp. 84–85.

21. Ibid., pp. 321–22.

22. Ibid., p. 225.

23. Ibid., p. 314.

24. Ibid., p. 333.

25. Ibid., p. 337.

26. Ibid., p. 339.

27. Lapham, "The Energy Debacle," p. 64.

28. Energy Policy Project of the Ford Foundation, *A Time To Choose* (Cambridge, Mass.: Ballinger, 1974), p. 325.

29. Ibid., p. 326.

30. Ibid.

31. Lapham, "The Energy Debacle," p. 67.

32. Ibid., p. 68.

33. Ibid., p. 59.

34. Ibid.

35. Ibid.

Chapter 5

1. Whitaker, *Striking a Balance,* p. 106.

2. William E. Simon, *A Time for Truth* (New York: McGraw-Hill, 1978), p. 79.

3. Ibid.

4. Ibid., p. 80.

5. Whitaker, *Striking a Balance,* p. 93.

6. Casey Bukro, "It's Dirty Jerry vs. Mr. Clean: Ecologists," *Chicago Tribune,* October 28, 1976.

7. "Carter Wins Praise of Environmentalists for Record So Far," *Wall Street Journal,* December 21, 1978.

8. Lapham, "The Energy Debacle," p. 70.

9. Ibid., p. 61.

10. Jimmy Carter, "Remarks of the President to the American People on the Energy Problems," The White House, April 18, 1977.

11. Jimmy Carter, "Remarks of the President to a Joint Session of Congress," The White House, April 20, 1977.

12. Jimmy Carter, "Remarks of the President to the American People on the Energy Problem."

13. "Carter Rips Oil-Gas Energy 'Profiteers,'" *Chicago Daily News,* October 13, 1977.

14. National Energy Act (Washington, D.C.: U.S. Government Printing Office, 1977), pp. 6–7.

15. Richard Rhodes, "Prodigy of Energy," *Quest/78,* December-January 1978, p. 22.

16. Ibid., p. 104.

17. Amory Lovins, "A Light on the Soft Path," *Sun! A Handbook for the Solar Decade* (San Francisco: Friends of the Earth, 1978), p. 39.

18. "'Invisible Crisis' Faces U.S., Schlesinger Warns," *Chicago Tribune,* October 2, 1977. See also Robert Young, "Carter Energy Team's in Washington—and Yet Far Away," *Chicago Tribune,* July 8, 1977.

19. *Access to Energy,* October 1978, p. 1.

20. Denis Hayes, *Energy: The Case for Conservation* (Washington, D.C.: Worldwatch Institute, 1976), p. 7.

21. H. Peter Metzger, "The Coercive Utopians: Their Hidden Agenda," *The Denver Post,* April 30, 1978.

22. Ibid.

23. Ibid.

24. Ibid.

25. "Energy Conservation, Alternative Sources Pushed by California," *Wall Street Journal,* September 12, 1978.

Chapter 6

1. W. Philip Gramm, "The Economics of the Energy Crisis," in David J. Theroux, ed., *The Energy Crisis: Government and the Economy* (San Francisco: Cato Institute, 1978), p. 27.

2. Fred H. Schmidt and David Bodansky, *The Energy Controversy: The Fight Over Nuclear Power* (San Francisco: Albion, 1976), p. 7.

3. Quoted in *Access to Energy,* July 1, 1976.

4. Richard J. Gonzalez, "Energy and Human Welfare," in Holt Ashley, Richard L. Rudman, Christopher Whipple, eds., *Energy and the Environment: A Risk Benefit Approach,* (Elmsford, N.Y.: Pergamon Press, 1976), p. 278.

5. John U. Nef, "An Early Energy Crisis and Its Consequences," *Scientific American,* November 1977, p. 140.

6. Ibid., p. 141.

7. Ibid., p. 140.

8. Ibid., p. 147.

9. Janet Fay Jester and William A. Jester, "Energy: Past, Present and Future," Pennsylvania State University, mimeographed paper, June, 1977, p. 9.

10. Edward J. Mitchell, "Implications of Zero Energy Growth in U.S.," *Purdue Energy Conference of 1977* (West Lafayette, Ind.: Purdue University Energy Engineering Center, 1977), p. 116.

11. Ibid., p. 116–17.

12. Schmidt and Bodansky, *The Energy Controversy*, p. 11.

13. Andrew L. Simon, *Energy Resources* (Elmsford, N.Y.: Pergamon Press, Inc., 1975), pp. 2–3.

14. U.S. Department of the Interior, *Energy Perspectives* (Washington, D.C.: U.S. Government Printing Office, 1975), p. 34, and Energy Information Administration, *Annual Report to Congress, Volume III, 1977* (Washington, D.C.: U.S. Government Printing Office, 1978), p. 5.

15. Gonzalez, "Energy and Human Welfare," pp. 281–83.

16. Ibid., pp. 282–83.

17. H. R. Linden, "Energy Policy and Resource Development," *Purdue Energy Conference of 1975*, pp. 9–11.

18. Energy Information Administration, *Annual Report to Congress, Volume III, 1977*, p. 5.

19. Ibid., p. 3.

20. Ibid., p. 9.

Chapter 7

1. Ben J. Wattenberg, "Out of Gas? Man's Wits Will Save Us," *Chicago Daily News*, March 31, 1975.

2. Edward J. Mitchell, *U.S. Energy Policy*, p. 5.

3. Ibid.

4. Jimmy Carter, "Remarks of the President to the American People on the Energy Problem."

5. Robert Young, "The Credibility Gap—Is the Crunch Really Coming?" *Chicago Tribune*, October 2, 1978.

6. John E. Swearingen, "Energy Facts and Fallacies," address to Young Presidents Club, Chicago, April 11, 1978.

7. "Enough Oil Until 2068 A.D.," *Chicago Tribune*, October 16, 1978.

8. Ibid.

9. "Forecast of Shortage of Soviet Oil in '80s Is Confirmed by CIA," *Wall Street Journal,* March 22, 1979.

10. "New Mideast War? One Saudi Believes That the Odds Favor It," *Barron's,* May 19, 1975.

11. Paul Scott, "Silencing Energy Dissent," *The Wanderer,* September 29, 1977.

12. Ibid.

13. "The Memory Hole," *Wall Street Journal,* April 4, 1978.

14. Amory B. Lovins, *World Energy Strategies: Facts, Issues and Options* (New York: Friends of the Earth, 1975), p. 84.

15. Harrison Brown, "Energy in our Future," *Annual Review of Energy, Volume 1* (Palo Alto, Calif.: Annual Reviews, Inc., 1976), p. 20.

16. Herman Kahn, William Brown and Leon Martel, *The Next 200 Years* (New York: Morrow, 1976), p. 63.

17. John E. Swearingen, "Energy Facts and Fallacies."

18. "Oil Reserve World's Biggest: Mexico," *Chicago Tribune,* September 1, 1978.

19. Ronald Yates, "Mexico Is Sitting on a Vast Sea of Oil, Experts Say," *Chicago Tribune,* November 19, 1978.

20. Stephen Rattien and David Eaton, "Oil Shale: The Prospects and Problems of an Emerging Industry," *Annual Review of Energy, Volume 1,* pp. 184–85.

21. Paul Scott, "Questionable Energy Assumptions," *The Wanderer,* May 19, 1977.

22. Carlos Henkel and Robert Poole, Jr., "How To Break Up OPEC," *The Freeman,* October 1978, pp. 584–85.

23. Energy Research and Development Administration, *Energy from Coal* (Washington, D.C.: U.S. Government Printing Office, 1976), p. 11-3.

24. Federal Energy Resources Council, "Uranium Reserves, Resources and Production," June 15, 1976.

25. Paul Kruger, "Geothermal Energy," *Annual Review of Energy, Volume 1,* p. 159.

26. R. F. Post, "Nuclear Fusion," *Annual Review of Energy, Volume I,* pp. 213–14.

27. Science and Public Policy Program, University of Oklahoma, *Energy Alternatives: A Comparative Analysis* (Washington, D.C.: U.S. Government Printing Office, 1975), p. 11-3.

28. Ibid., p. 11-15.

29. Ibid., p. 9-3.

30. Ibid., p. 9-15.

31. Ibid., pp. 11-19, 11-20.

32. Ibid., pp. 10-1-10-5.

33. Ibid., p. 11-25.

34. Lovins, *World Energy Strategies*, p. 8.

35. John Maddox, *Beyond the Energy Crisis* (New York: McGraw-Hill, 1975), p. 49.

Chapter 8

1. Fred Hoyle, *Energy or Extinction?, The Case for Nuclear Energy*, (Salem, N.H.: Heinemann Educational Books Inc., 1977), p. 39.

2. Ibid.

3. Thomas F. P. Sullivan, ed., *Energy Reference Handbook* (Washington, D.C.: Government Institutes, Inc., 1977), pp. 316-17.

4. George H. Galloway, "Time, Technology and Price Incentives Will Have Major Impact on the Volume of Future U.S. Reserves and Production," *SpaN* (Standard Oil Company of Indiana, Chicago), Number 3, 1977, p. 14.

5. Arlen J. Large, "NASA Finds Many Buyers for Satellite Pictures and Plans New, More Revealing Landsats in '80s," *Wall Street Journal*, April 5, 1979.

6. Peter Nulty, "When We'll Start Running Out of Oil," *Fortune*, October 1977, p. 246.

7. "Shell's Wonderful World of Oil: Offshore Drilling," Shell Oil Company, n.d.

8. "Continental Oil, U.S. Project," *Wall Street Journal*, April 10, 1978.

9. "Shale Oil—All Dressed Up but No Place to Go," *Developing the Many Sources of Energy* (Houston, Texas: Public Affairs Department, Exxon Company, USA, n.d.).

10. Marc W. Goldsmith, Ian A. Forbes, Joe C. Turnage, Stacy V. Weaver and Alan R. Forbes, *New Energy Sources: Dreams and Promises* (Waltham, Mass.: Energy Research Group, 1976), p. 4.

11. "Unit of Occidental Petroleum Gets Job Involving Oil Share," *Wall Street Journal*, October 6, 1977.

12. "Two Promising Oil Shale Processes," *New York Times*, July 27, 1977.

13. "Oil 'Shortages'—a Chicago Solution," *Chicago Tribune*, May 9, 1978.

14. Science and Public Policy Program, *Energy Alternatives*, pp. 4-11.

15. Privately circulated paper of major research organization.

16. Henkel and Poole, "How To Break Up OPEC," p. 585.

17. Office of Technology Assessment, Congress of the United States, *Gas Potential from Devonian Shales of the Appalachian Basin* (Washington, D.C.: U.S. Government Printing Office, 1977), p. 3

18. Science and Public Policy Program, *Energy Alternatives,* p. 1-31.

19. Ibid., p. 1-47.

20. Ibid., p. 1-44.

21. Goldsmith *et al., New Energy Sources: Dreams and Promises,* pp. 5–6.

22. "Uranium: Energy for the Future," Atomic Industrial Forum, Inc., Washington, D.C.

23. "How Nuclear Plants Work," Atomic Industrial Forum, Inc., Washington, D.C.

24. "Recycling Nuclear Fuel," Atomic Industrial Forum, Inc., Washington, D.C.

25. "Reprocessing, Recycle and the Breeder," Atomic Industrial Forum, Inc., Washington, D.C.

26. Science and Public Policy Program, *Energy Alternatives,* p. 6-42.

27. Ibid., p. 6–34.

28. "Recycling Nuclear Fuel."

29. Science and Public Policy Program, *Energy Alternatives,* p. 6-58.

30. Ibid., p. 7–1.

31. "Does Fusion Hold Key to U.S. Energy Future?" *U.S. News & World Report,* August 28, 1978.

32. James Pearre, "KMS a 'David' in Fusion Contest," *Chicago Tribune,* June 15, 1975.

33. Science and Public Policy Program, *Energy Alternatives,* p. 8-5.

34. Ibid., p. 8-13.

35. Ibid., p. 8-5.

36. Ibid., p. 8-14

37. Direct-mail letter, Friends of the Earth, n.d.

38. Lovins, "Energy Strategy," p. 81.

Chapter 9

1. *Solar Energy: A Status Report* (Washington, D.C.: U.S. Department of Energy, 1978), p. 24.

2. Goldsmith *et al., New Energy Sources,* p. 17.

3. Ibid., p. 16.

4. Ibid.

5. Ibid.

6. Ibid., pp. 10–11.

7. Ibid., p. 11.

8. Ibid., p. 14.

9. Ibid., p. 15.

10. Ibid., pp.12–13.

11. *Solar Energy: A Status Report,* p. 65.

12. Goldsmith *et al.,New Energy Sources,* p. 7.

13. Science and Public Policy Program, *Energy Alternatives,* p. 11-22.

14. Goldsmith *et al., New Energy Sources,* p. 10.

15. Lovins, "Energy Strategy," pp. 80–81.

16. Ibid., p. 82.

17. Ralph Lapp, "Summary of 'A Critique of Amory Lovins' Article,'" in Charles B. Yulish, ed., *Soft vs. Hard Energy Paths* (New York: Charles Yulish Associates, 1977), p. 61.

18. Lovins, "Energy Strategy," p. 82.

19. Bruce Adkins, "Misdirections in Energy Strategy," *Soft vs. Hard Energy Paths,* p. 4.

20. Aden Meinel and Marjorie Meinel, "'Soft' Energy Paths—Reality and Illusion," *Soft vs. Hard Energy Paths,* p. 70.

21. Sheldon H. Butt, "A Solar View of the Soft Path," *Soft vs. Hard Energy Paths,* pp. 16, 28.

Chapter 10

1. Edward J. Mitchell, *The Energy Dilemma: Which Way Out?* (Washington, D.C.: American Enterprise Institute for Public Policy Research, 1975), p. 2.

2. Ibid.

3. Energy Research and Development Administration, *Market Oriented Program Planning Study (MOPPS), Integrated Summary, Vol. 1, Final Report,* December 1977.

4. Stephen Rattien and David Eaton, "Oil Shale: The Prospects and Problems of an Emerging Energy Industry," *Annual Review of Energy, Volume 1,* p. 194.

5. *Market Oriented Program Planning Study.*

6. Henkel and Poole, "How To Break Up OPEC," pp. 584–85.

7. Energy Information Administration, *Annual Report to Con-*

gress, *Volume II—1977* (Washington, D.C.: U.S. Government Printing Office, 1978), table 8-7, p. 183.

8. *Statistical Data of the Uranium Industry,* GJO-100 (78), U.S. Department of Energy, January 1, 1978.

9. Direct-mail letter, Friends of the Earth, n.d.

10. Lovins, "Energy Strategy," p. 81.

11. Ibid., p. 83.

12. Frederick H. Morse and Melvin K. Simmons, "Solar Energy," *Annual Review of Energy, Volume 1,* p. 143.

13. Lovins, "Energy Strategy," p. 69.

14. Ibid., p. 87.

15. Ibid.

16. Ibid., p. 66.

17. Ian A. Forbes, "The Economics of Amory Lovins' Soft Path," Energy Research Group, Inc., 400-1 Totten Pond Road, Waltham, Mass., mimeographed report, pp. 8–10.

18. Ibid., p. 11.

19. Ibid., p. 10.

20. Ibid., pp. 1–2.

21. Ibid., p. 6. Some of the cost results shown here were calculated from data developed by Forbes in dollars per barrel per day.

22. Daniel W. Kane, "Comments on Article by Amory Lovins," *Soft vs. Hard Energy Paths,* pp. 51–53.

23. Ibid., pp. 54–55.

24. Energy Research Group, Inc., *Assessment of Alternate Technologies for Utility Baseload Generating Capacity in New England* (Waltham, Mass.: Energy Research Group, Inc., 1979), p. 148. Power costs reflect base construction costs which do not include contingency, interest and escalation during construction and owner's costs.

25. Ibid., p. 151.

26. Peter J. Reilly, "Economics and Energy Requirements of Ethanol Production," Department of Chemical Engineering and Nuclear Engineering, Iowa State University, Ames, Iowa, mimeographed paper, pp. 1, 22.

27. "Solar Energy: An Introduction," Solar Lobby, 1028 Connecticut Avenue, N.W., Washington, D.C.

28. Congressman Mike McCormack, "Cause for Celebration: A Realistic Look at Solar Programs," in Milton R. Copulos, ed., *Energy Perspectives* (Washington, D.C.: Heritage Foundation, 1978), p. 118.

29. Quoted in Forbes, "The Economics of Amory Lovins' Soft Path," p. 3.

30. Presentation by Marjorie and Aden Meinel, University of

Arizona, at Public Forum in Helena, Montana, April 26-27, 1978, reprinted by Atomic Industrial Forum, Inc.

31. Meinel and Meinel, "Soft Energy Paths—Reality and Illusion," *Soft vs. Hard Energy Paths,* pp. 73-76.

32. Howard Odum, "Net Energy from the Sun," in Stephen Lyons, ed., *Sun! A Handbook for the Solar Decade* (San Francisco: Friends of the Earth, 1978), p. 207.

33. William Shurcliff, "Eight Untried Inventions and a Plea," *Sun! A Handbook for the Solar Decade,* p. 270.

34. Schumacher, *Small is Beautiful,* p. 3.

35. Ibid., p. 9.

36. Amory B. Lovins, *Soft Energy Paths* (San Francisco: Friends of the Earth, 1977), pp. 70-72.

37. Stephen Lyons, ed., "Introduction," *Sun! A Handbook for the Solar Decade,* pp. 2, 4.

38. "Solar Energy: An Introduction," p. 34.

39. "Solar Energy Programs and Goals for Year 2000 Are Proposed by Carter," *Wall Street Journal,* June 21, 1979.

40. Ibid.

41. "Solar Energy: An Introduction," p. 18.

Chapter 11

1. L. A. Sagan and A. A. Afifi, *Health and Economic Development II: Longevity* (Laxenburg, Austria; International Institute for Applied Systems Analysis, 1978), p. v.

2. James R. Dunn, *The Price of a Good Environment* (Louisville, Ky.: National Council for Environmental Balance, Inc.).

3. Ibid.

4. Quoted in Ibid.

5. *Access to Energy,* October 1977.

6. Dunn, *The Price of a Good Environment.*

7. Whitaker, *Striking a Balance,* p. 24.

8. Remarks of David E. Lilienthal at the Utility Regulatory Conference sponsored by Public Utilities Reports, Inc., Mayflower Hotel, Washington, D.C., October 4, 1978.

9. "The Environmental Complex: Part II," *Institution Analysis,* Heritage Foundation, Washington, D.C., p. 1.

10. "The Environmental Complex," *Institution Analysis,* Heritage Foundation, Washington, D.C., pp. 6-7.

11. Ibid., p. 8.

12. AMA Council on Scientific Affairs, "Health Evaluation of

Energy-Generating Sources," *Journal of the American Medical Association,* November 10, 1978, pp. 2193–94.

13. Energy Policy Project of the Ford Foundation, *A Time To Choose,* pp. 270–73.

14. H. Peter Metzger, "The Coercive Utopians: Their Hidden Agenda," address to the Alaska State Chamber of Commerce 19th Annual Convention, Sitka, Alaska, October 6, 1978, p. 8.

15. "Stalled Program of Leasing Coal Rights on U.S. Lands Could Begin by Mid-1980," *Wall Street Journal,* December 18, 1978.

16. Metzger, "The Coercive Utopians: Their Hidden Agenda," p. 9.

17. Ibid.

18. Ibid., pp. 9–10.

19. "U.S. To Resume Leasing Rights to Federal Coal," *Wall Street Journal,* June 5, 1979.

20. Letter, Carl E. Bagge, president of the National Coal Association, to Hon. Frank Gregg, director, Bureau of Land Management, Office of Coal Management, U.S. Department of the Interior, February 13, 1979, p. 4.

21. Margot Hornblower, "Oil Spills Do Little Real Harm to Wildlife or Man: Experts," *New York Post,* March 14, 1977.

22. "Oil Spills and Oil Reserves," *Wall Street Journal,* July 26, 1979.

23. "High-Stakes Gamble for Atlantic Oil," *U.S. News & World Report,* May 8, 1978, p. 30.

24. Natural Resources Defense Council, *Annual Report 1977/78,* pp. 9–10.

25. Neil Ulman, "Behind the Georges Bank Battle," *Wall Street Journal,* March 3, 1978.

26. "U.S. Aims to Boost Oil Lease Sales," *Chicago Tribune,* June 8, 1979.

27. William E. Blundell, "Industrialists Losing War of the Wilderness as Combat Rages On," *Wall Street Journal,* June 18, 1979.

28. Metzger, "The Coercive Utopians: Their Hidden Agenda," p. 14.

29. Walter J. Hickel, "Keeping Alaska on Ice," *Saturday Evening Post,* July-August 1979, p. 114.

30. "Where the Caribou Play," *Wall Street Journal,* June 5, 1979.

31. Margot Hornblower, "Wildlife Species Act Endangers Projects," *Chicago Daily News,* July 21, 1977.

32. "Three-inch fish torpedoes $119 Million Dam," *Chicago Tribune,* January 24, 1979.

33. "Struggle Over Seabrook Whale of a Controversy," *Nuclear*

Info, Atomic Industrial Forum, Washington, D.C., p. 4.

34. Milton R. Copulos, "Coal Conversion: Costs and Conflicts," *Backgrounder*, Heritage Foundation, Washington, D.C., September 22, 1977, p. 2.

35. John E. Swearingen, "Energy, Agriculture, and America's Future: Restoring the Natural Balance," address to National Association of County Agricultural Agents, Boise, Idaho, August 14, 1978.

36. William N. Poundstone, "SO_2: A Burning Issue," *Coal Mining & Processing*, November 1978, p. 54.

37. Ibid., pp. 54–56.

38. Irwin Tucker, "Presentation at EPA FGD Hearings," December 12, 1978, National Council for Environmental Balance, Louisville, Ky.

39. Eugene Guccione, "An Open Letter to President Carter," *Coal Mining & Processing*, January 1979, p. 9.

40. Eugene Guccione, "The SO_2 Emission Fraud," *Coal Mining & Processing*, November 1978, p. 9.

41. Ibid.

42. Poundstone, "SO_2: A Burning Issue," p. 56.

43. *Access to Energy*, October 1977.

44. Robert Young, "Pipeline Plan Stirs Debate," *Chicago Tribune*, March 28, 1979.

45. "Notable & Quotable," *Wall Street Journal*, June 18, 1979.

46. "No-Growth Muggers," *Great News*, P.O. Box 50179, Washington, D.C., p. 6.

47. Douglas Martin, "Environmental Risk From 'Renewable' Energy Sources?" *Wall Street Journal*, June 9, 1978.

48. Richard Singer and Thomas C. Roberts, "Land Use Requirements for Five Energy Alternatives," *Reference Documents on Energy-Related Issues* (La Grange Park, Ill.: American Nuclear Society).

49. Kathryn A. Lawrence, *A Review of the Environmental Effects and Benefits of Selected Solar Energy Technologies* (Springfield, Va.: National Technical Information Service, 1978), p. 3.

50. Martin, "Environmental Risk From 'Renewable' Energy Sources?"

51. Ken Bossong, "Hazards of Solar Energy," Citizens' Energy Project, Washington, D.C., p. 1.

52. Christopher D. Stone, *Should Trees Have Standing?* (Los Altos, Calif.: William Kaufman, 1974), p. 9.

53. Ibid., p. 18.

54. Ibid., p. 19.

55. Ibid., p. 34.

56. Ibid., pp. 40, 43, 49, 52.

57. Ibid., p. xiii.

58. Ibid., p. xiv.

59. Ibid., p. 57.

60. Ibid., p. 73.

61. Ibid., p. 93.

62. Ibid., p. 87.

63. Margaret N. Maxey, "Energy, Society and the Environment: Conflict or Compromise?" *Ethics and Energy* (Washington, D.C.: Edison Electric Institute, 1979), p. 37.

64. Ibid., pp. 37–38.

65. "EPA Chief Will Direct New Panel to Monitor Effects of Regulations," *Wall Street Journal,* November 1, 1978.

66. "Proposed Energy Board Would Expedite As Many As 75 Projects at Any One Times," *Wall Street Journal,* July 20, 1979.

67. Robert J. Wagman and Sheldon D. Engelmayer, "Broken: One Fuel Vow," *Desert News* (Salt Lake City), July 25, 1979.

68. Ibid.

Chapter 12

1. Walter Oi, "Safety at any Price?" *Regulation* (American Enterprise Institute), November-December 1977, p. 17.

2. Robert Enstad, "Don't Put Your Number in This Book by Accident," *Chicago Tribune,* May 25, 1979.

3. Herbert Inhaber, *Risk of Energy Production* (Ottawa, Ontario, Canada: Atomic Energy Control Board, 1978), p. v.

4. Ibid., p. 39.

5. "Reactor Safety Study: An Assessment of Accident Risks in U.S. Commercial Nuclear Power Plants, Executive Summary." (WASH-1400), Nuclear Regulatory Commission, Washington, D.C., 1975.

6. Bill Raspberry, "Nuclear Power Is Safe, but . . . ," *Chicago Tribune,* April 10, 1979.

7. Quoted in Bertram Wolfe, "Some Thoughts on New Energy Sources," *Nuclear News,* May 1976.

8. H. Peter Metzger, "Anti-Nuclear Initiatives—Hoax on Public," *Rocky Mountain News,* May 9, 1976.

9. Ibid.

10. Jacque Srouji, *Critical Mass: Nuclear Power, the Alternative to Energy Famine* (Nashville, Tenn.: Aurora, 1977), p. 47.

11. Petr Beckmann, *The Health Hazards of NOT Going Nuclear* (Boulder, Colo.: Golem Press, 1976), pp. 18–19.

12. Metzger, "Anti-Nuclear Initiatives."

13. Peter Cowen, "Questioning the Safety of Nuclear Power Plants," *Boston Sunday Globe,* August 22, 1976.

14. "What the Anti-NUKE Alliances are Up To: Signals from the Other Side," *Great News,* January 1979.

15. Metzger, "Anti-Nuclear Initiatives."

16. Ibid.

17. "The Plowboy Interview: David Brower, a Friend of the Earth."

18. "Nuclear Power and the Sierra Club," Sierra Club, November 1977.

19. Thomas J. Connolly, "Implications of the California Nuclear Power Plant Initiatives," testimony before the California State Assembly Committee on Energy and Diminishing Materials, Sacramento, Calif., December 2, 1975. Connolly's mention of a "phony story" in the *Chicago Tribune* is a reference to an October 1975 front-page story which described in detail how a U.S. spy satellite photographed the results of the explosion of a Russian nuclear reactor, showing a large crater, fallen trees, human bodies and no signs of life in the area. The story turned out to be an outright hoax fed to the *Tribune* by antinuclear activists.

20. Helen Caldicott, "At the Crossroads," *New Age,* December 1977, p. 1.

21. John Berendt, "A Day in the Life of a Nuclear Power Plant," *Cosmopolitan,* January 1978, p. 145.

22. "Warning: Opponents of Nuclear Power May Be Dangerous to Your Future," reprinted by Florida Power & Light Company from a July 1977 *Wall Street Journal* advertisement by Dresser Industries, Dallas, Texas.

23. John W. Gofman, "A Small Affidavit with Big Implications," Committee for Nuclear Responsibility, Inc., P.O. Box 11207, San Francisco, Calif., June 5, 1978.

24. "Background Data on Radiation," mimeographed paper by Norman M. Cole, Jr., Lorton, Va., March 8, 1977.

25. "Surprising Contradictions Found Among Japanese A-Bomb Survivors," *Cincinnati Post,* October 20, 1977.

26. Bernard L. Cohen, "Impacts of the Nuclear Energy Industry on Human Health and Safety," *American Scientist,* September-October, 1976, pp. 550–51.

27. *Nuclear Power and the Environment* (Hinsdale, Ill.: American Nuclear Society, 1976), p. 30.

28. Daniel F. Ford, "Are We Going Overboard on Nuclear Energy?" *Boston Sunday Globe,* May 29, 1977.

29. Beckmann, *The Health Hazards of NOT Going Nuclear,* p. 46.

30. "The Brookhaven Reports," *Background Info,* Atomic Industrial Forum, Inc., Washington, D.C.

31. Aaron Latham, "Hollywood vs. Harrisburg," *Esquire,* May 22, 1979.

32. Quoted in Daniel F. Ford and Henry W. Kendall, "The Nuclear Power Issue: An Overview," Union of Concerned Scientists, Cambridge, Mass.

33. "Report Blames Operators for 3 Mile Accident," *Chicago Tribune,* August 3, 1979.

34. "Summary and Discussion of Findings from: Population Dose and Health Impact of the Accident at the Three Mile Island Nuclear Station," Bureau of Radiological Health, Rockville, Md.

35. Ibid.

36. Ibid.

37. Ibid.

38. "See No Risk for Unborn," *Chicago Tribune,* April 13, 1979.

39. Samuel McCracken, "The Harrisburg Syndrome," *Commentary,* June 1979, p. 27.

40. "Three Mile Island Bubble Posed No Blast Risk: NRC," *Chicago Tribune,* April 2, 1979.

41. Edward Teller, "Dr. Teller Defends Nuclear Power," *Chicago Tribune,* April 11, 1979.

42. William Waldegrave, "The Nuclear Case that Must Be Seen to Be Won," *The Guardian* (London), July 2, 1979.

43. "Reactor Safety Study," p. 10.

44. "NRC Statement on Risk Assessment and the Reactor Safety Study Report (WASH-1400) in Light of the Risk Assessment Review Group Report," Nuclear Regulatory Commission, Washington, D.C., January 18, 1979.

45. Cohen, "Impacts of the Nuclear Energy Industry on Human Health and Safety," p. 558.

46. "Waste Management Studies: Conclusions and References," *Background Info,* Atomic Industrial Forum, Inc., Washington, D.C., August 1978.

47. Letter, Bernard L. Cohen, professor of physics and director of Scaife Nuclear Physics Laboratory, n.d.

48. Petr Beckmann, *Nuclear Proliferation: How to Blunder Into Promoting It* (Boulder, Colo.: Golem Press, 1977), p. 5.

49. "Scientists Say Reactor Ends Threat of Spreading Bombs,"

Chicago Tribune, February 28, 1978.

50. Beckmann, *Nuclear Proliferation: How To Blunder Into Promoting It*, p. 4.

51. Samuel McCracken, "The War Against the Atom," *Commentary*, September 1977, p. 35.

52. "Scientists' Statement on Safety of Nuclear Energy," Scientists and Engineers for Secure Energy, New York, pp. 1–2.

53. Herbert Inhaber, "Risk of Energy Production," *Energy Perspectives*, p. 223.

54. Ibid., pp. 223–24.

55. Herbert Inhaber, *Risk of Energy Production* (Ottawa, Ontario, Canada: Atomic Energy Control Board, 1978), p. 50.

56. Suzanne Weaver, "The Passionate Risk Debate," *Wall Street Journal*, April 24, 1979.

57. Letter, John P. Holdren to the *Wall Street Journal*, May 7, 1979.

58. Letter, Herbert Inhaber to the *Wall Street Journal*, June 18, 1979. Inhaber summed up his letter by stating, "I regret that Prof. Holdren has chosen to communicate only with my employers or newspapers, not directly with me. If this is what passes for scientific communications in the United States, I trust that it will not penetrate north of the border."

59. *Annual Report 1977/78*, Natural Resources Defense Council, New York, N.Y.

60. Quoted in *Proceedings of the Purdue Energy Conference of 1977* held at Purdue University, West Lafayette, Ind., April 29–30, 1977.

61. "CEQ Would Halt Nuclear; Other Spokesmen Disagree," *Nuclear News*, November 1977, p. 36.

62. Metzger, "The Coercive Utopians: Their Hidden Agenda."

63. Senator James A. McClure, "United States Nuclear Power Isolationism: A Case Study in the Abdication of World Leadership," address to the Commonwealth Club of California, San Francisco, Calif., February 16, 1979.

64. Jon Margolis, "Huge Rally on Steps of Capitol," *Chicago Tribune*, May 7, 1979.

65. Ira R. Allen, "Carter Backs Nuclear Power," *Cleveland Press*, May 8, 1979.

66. Charles Bartlett, "Carter Must Take Nuclear Energy Lead," *Chicago Daily News*, July 15, 1977.

67. Margolis, "Huge Rally."

68. McClure, "United States Nuclear Power Isolationism."

69. Walter Meyer, Myron R. Rollins and Raymond K. Williams, "Estimate of the Economic Effects of a Five Year National Nuclear Power Moratorium," presented at the Workshop on the Nuclear Debate: Basic Issue for 1976, sponsored by the Atomic Industrial Forum, October 29, 1975.

70. "Nuclear Cost/Benefits Weighed by NRC Staff," *Nuclear Info,* Atomic Industrial Forum, Washington, D.C., December 1978.

Chapter 13

1. D. T. Armentano, "Interventionism in the Petroleum Industry," *The Energy Crisis,* p. 3.

2. Edward J. Mitchell, "The Implications of Zero Energy Growth in the United States," *Proceedings of the Purdue Energy Conference of 1977,* p. 117.

3. Armentano, "Interventionism in the Petroleum Industry," pp. 3–4.

4. William H. Peterson, *The Question of Governmental Oil Import Restrictions* (Washington, D.C.: American Enterprise Association, 1959), pp. 10–12.

5. Ibid., p. 12.

6. Ibid., p. 13.

7. Mitchell, *U.S. Energy Policy,* p. 29.

8. Armentano, "Interventionism in the Petroleum Industry," p. 4.

9. Mitchell, *U.S. Energy Policy,* pp. 35–36.

10. Peterson: *The Question of Governmental Oil Import Restrictions,* p. 36.

11. Ibid., p. 37.

12. J. Clayburn La Force, *The Energy Crisis: The Moral Equivalent of Bamboozle* (Los Angeles: International Institute for Economic Research, 1978), p. 10.

13. Mitchell, *U.S. Energy Policy,* p. 38.

14. Peterson, *The Question of Governmental Oil Import Restrictions,* p. 87.

15. C. John Miller, "American Oil: Our Bridge to the Future," *Imprimis,* Hillsdale College, Hillsdale, Michigan, May 1978.

16. David A. Loehwing, "Man-Made Shortage," *Barron's,* October 18, 1971.

17. Mitchell, *U.S. Energy Policy,* pp. 63–64.

18. Energy Information Administration, *Annual Report to Congress, Statistics and Trends of Energy Supply, Demand and Prices, Volume III, 1977* (Washington, D.C.: U.S. Government Printing

Office, 1978), pp. 19, 105. Dollar figures for electricity represent the weighted average monthly bill of residential consumers for 500 kilowatt-hours.

19. Ibid. Figures for electricity are expressed in 1972 dollars using GNP implied price deflators given in source.

20. Armentano, "Interventionism in the Petroleum Industry," p. 5.

21. Peterson, *The Question of Governmental Oil Import Restrictions*, pp. 50–51.

22. Armentano, "Interventionism in the Petroleum Industry," p. 5.

23. Mitchell, *U.S. Energy Policy*, pp. 45–46.

24. Simon, *A Time For Truth*, p. 73.

25. Ibid., pp. 79–80.

26. "Guest Editorial," *Wall Street Journal,* November 8, 1977.

27. Robert S. Pindyck, "Prices and Shortages: Policy Options for the Natural Gas Industry," *Options for U.S. Energy Policy* (San Francisco, Calif.: Institute for Contemporary Studies, 1977), p. 170.

28. John J. McKetta, "Why the U.S. Will Not Achieve Energy Self-Sufficiency," lecture presented at the University of Louisville, April 11, 1979, distributed by National Council for Environmental Balance, Louisville, Ky., p. 7.

29. John J. McKetta, "The U.S. Energy Problem—America's Achilles Heel," lecture presented at Pennsylvania State University, March 21, 1978, distributed by National Council for Environmental Balance, Louisville, Ky., pp. 6–7.

30. McKetta, "Why the U.S. Will Not Achieve Energy Self-Sufficiency," p. 8.

31. "Restraints Restrict Return," *Energy,* vol. XIV, no. 3, 1978, Citibank, New York.

32. Melvin R. Laird, *Energy—A Crisis in Public Policy* (Washington, D.C.: American Enterprise Institute for Public Policy Research, 1977), p. 7.

33. Milton Friedman, "The Energy Crisis: A Humane Solution," *The Energy Crisis, Government and the Economy,* p. 16.

34. Quoted La Force, *The Energy Crisis*, p. 22.

35. "Windfall Profits or Windfall Tax?" Mobil advertisement, *Wall Street Journal,* April 17, 1979.

36. "Ecology Groups to Ask Decontrol of Oil Prices," *Wall Street Journal,* March 21, 1979.

37. Arthur Siddon, "Oil Profit a Windfall for Carter," *Chicago Tribune,* July 31, 1979.

38. "Nader Backs Consumer Boycott," *Chicago Tribune,* June 28, 1979.

39. "Congress Gets Carter Plan for Oil Tax, President Warns Against Big Loopholes," *Wall Street Journal,* April 27, 1979.

40. "Some Senators Skeptical of Carter Plans for Spending Oil Windfall Tax Proceeds," *Wall Street Journal,* August 1, 1979.

41. Paul L. Joskow and Robert S. Pindyck, "Those Subsidized Energy Schemes," *Wall Street Journal,* July 2, 1979.

42. "Nader Backs Consumer Boycott."

43. Paul W. McCracken, "Carter's Confidence Speech," *Wall Street Journal,* August 1, 1979.

44. Ibid., p. 23.

Chapter 14

1. Meinel and Meinel, "'Soft' Energy Paths—Reality and Illusion," *Soft vs. Hard Energy Paths,* p. 72.

2. Sam S. Schurr, "Energy, Economic Growth, and Human Welfare," *Ethics and Energy* (Washington, D.C.: Edison Electric Institute, 1979), pp. 20–21.

3. Ibid., p. 17.

4. Ian A. Forbes and Joe C. Turnage, *Exclusive Paths and Difficult Choices: An Analysis of Hard, Soft and Moderate Energy Paths* (Framingham, Mass.: Energy Research Group, Inc., 1977), p. 26.

5. Schurr, "Energy, Economic Growth, and Human Welfare," p. 18.

6. Forbes and Turnage, *Exclusive Paths,* p. 25.

7. Schurr, "Energy, Economic Growth and Human Welfare," p. 19.

8. H. R. Linden, "Energy Policy and Resource Development," *Purdue Energy Conference of 1975* (West Lafayette, Ind.: Purdue University Energy Engineering Center, 1975), pp. 8–9.

9. Hedrick Smith, "President Evokes Spirit of '76," *International Herald Tribune,* July 16, 1977.

10. "Remarks of the President to the American People on the Energy Problem."

11. Young, "Carter Energy Team's in Washington—and Yet Far Away."

12. "Energy: Will Americans Pay the Price?" *U.S. News & World Report,* May 2, 1977.

13. Lovins, "Energy Strategy," p. 72.

14. Alfred L. Malabre, Jr., "Despite the Dollar's Decline, U.S.

Retains Top Living Standard Among Major Nations," *Wall Street Journal,* May 1, 1979.

15. Joy Dunkerley, ed., *International Comparisons of Energy Consumption* (Washington, D.C.: Resources for the Future, 1978), pp. xxvi–xxvii.

16. Peter J. Brennan, "The Soft Path, The Yellow Brick Road and Pie in the Sky," *Soft vs. Hard Energy Paths,* pp. 8–9.

17. *Access to Energy,* October 1976.

18. Meinel and Meinel, "Energy/GNP Trajectories: The Relationship Between Economic Growth and Energy Consumption," *Ethics and Energy,* pp. 8–11.

19. Denis Hayes, *Energy: The Case for Conservation* (Washington, D.C.: Worldwatch Institute, 1976), p. 7.

20. Council on Environmental Quality, *The Good News About Energy* (Washington, D.C.: U.S. Government Printing Office, 1979), p. iv.

21. Schurr, "Energy, Economic Growth, and Human Welfare," pp. 17–18.

22. Hoyle, *Energy or Extinction?* pp. 24–25.

23. Ian A. Forbes, *Energy Strategy: Not What But How* (Framingham, Mass.: Energy Research Group, Inc., 1977), p. 15.

24. Council on Environmental Quality, *The Good News About Energy,* pp. iii–viii.

25. Mitchell, "The Implications of Zero Energy Growth in the United States," *Purdue Energy Conference of 1977,* p. 120.

26. Schumacher, *Small Is Beautiful,* p. 68.

27. Lovins, *World Energy Strategies,* p. 103.

28. Lovins, *Soft Energy Paths,* p. 165.

29. "Transportation," *Progress As If Survival Mattered,* p. 139.

30. Quoted in B. Bruce-Briggs, *The War Against the Automobile* (New York: E.P. Dutton, 1975), p. 191.

31. Ibid.

32. Ibid.

33. "Mileage Rules to Cost $80 Billion: Auto Exec," *Chicago Tribune,* March 15, 1979.

34. "Commerce Unit Study Backs Auto Firms on Cost Claims for Fuel-Economy," *Wall Street Journal,* May 14, 1979.

35. "A Peanut Butter Car?" *Wall Street Journal,* December 29, 1978.

36. Raymond E. Goodson, "Energy Conservation in Transportation," *Purdue Energy Conference of 1975,* p. 50.

37. Citizens' Advisory Committee on Environmental Quality, *Citi-*

zen Action Guide to Energy Conservation (Washington, D.C.: U.S. Government Printing Office, 1973), pp. 21-22.

38. U.S. Environmental Protection Agency, *Bicycle Transportation* (Washington, D.C.: U.S. Government Printing Office, 1974), pp. 15-22.

39. "HUD Energy Use Standards Could Alter Building Design," *Chicago Tribune,* March 19, 1978.

40. Bernard J. Frieden, *The Environmental Protection Hustle* (Cambridge, Mass.: The MIT Press, 1979), pp. 4, 9, 171.

41. "U.S. City Moves to Front in 'War to Save Energy,'" *International Herald Tribune,* July 11, 1979.

42. George Getschow, "Indoor Air Pollution Worries Experts As Buildings Are Sealed to Save Fuel," *Wall Street Journal,* August 15, 1979.

43. John P. Holdren, "Too Much Energy, Too Soon," *New York Times,* July 23, 1975.

44. Schurr, "Energy, Economic Growth and Human Welfare," pp. 19-20.

45. Meinel and Meinel, "Energy/GNP Trajectories: The Relationship Between Economic Growth and Energy Consumption," p. 12.

46. Ibid., p. 13.

47. Lovins, *World Energy Strategies,* p. 17.

48. James P. Gannon, "Some Americans Speak the Blues," *Wall Street Journal,* February 24, 1978.

49. Lovins, "Energy Strategy," p. 76.

50. John S. Steinhart, Mark E. Hanson, Rubin W. Gates, Carol C. DeWinkel, Kathleen Briody, Mark Thornsjo, and Stanley Kabala, "A Low Energy Scenario for the United States: 1975-2050," in Lon C. Ruedisili and Morris W. Firebaugh, eds., *Perspectives on Energy* (New York: Oxford University Press, 1978), p. 556.

51. Ibid., pp. 564-65.

52. "1984 Is Just Around the Corner," *The Mindszenty Report,* February 1978, p. 3.

53. Richard Spong, "Energy Approach Juvenile," *Miami Herald,* September 17, 1977.

54. William H. Riker, "The Ideology of A Time to Choose," *No Time to Confuse* (San Francisco: Institute for Contemporary Studies, 1975), pp. 155-56.

Chapter 15

1. Lovins, "Energy Strategy," p. 93.

2. Lovins, *World Energy Strategies,* p. 127.

3. Robin Clarke, ed., *Notes for the Future, An Alternative History of the Past Decade* (London: Thames and Hudson Ltd., 1975), p. 166.

4. Ernest Callenbach, *Ecotopia* (Berkeley, Calif.: Banyan Tree Books, 1975).

5. Ibid., pp. 43–46.

6. Ibid., p. 18.

7. Ibid., p, 86.

8. Ibid., pp. 90–93.

9. Ibid., pp. 27–28.

10. Ibid., pp. 40–41.

11. Ibid., p. 21.

12. Ibid., p. 29.

13. Ibid., p. 58.

14. Ibid., p. 55.

15. Ibid., p. 30.

16. Ibid., p. 32.

17. Ibid., p. 78.

18. Ibid., p. 34.

19. Ibid., p. 30.

20. Ibid., p. 22.

21. Ibid., pp. 61–63.

22. Ibid., p. 124.

23. Ibid., pp. 102–6.

24. Ibid., p. 147.

25. Ibid., p. 125.

26. Ibid., p. 3.

27. Ibid., p. 99.

28. Ibid., p. 86.

29. Ibid.

30. Ibid., pp. 164–67.

31. Ibid., back cover.

32. H. Peter Metzger, "Nuclear Roadblocks Mounting," *Rocky Mountain News,* September 19, 1976.

33. *Nuclear Info,* Atomic Industrial Forum, Washington, D.C., June 1978, p. 4.

34. John and Sally Seymour, "Self-sufficiency: What Is It? Why Do It?" *Notes for the Future, An Alternative History of the Past Decade,* pp. 199–201.

35. "Two Crises Compared," *Chicago Tribune,* June 28, 1979.

36. Brennan, "The Soft Path, The Yellow Brick Road and Pie in the Sky," *Soft vs. Hard Energy Paths,* p. 7.

37. Myron Kandel and Philip Greer, "Need for Energy Policy Critical, Study Warns," *Chicago Tribune,* May 28, 1978.

38. Milton R. Copulos, *Closing the Nuclear Option* (Washington, D.C.: Heritage Foundation, 1978), p. 5.

39. Metzger, "Anti-Nuclear Initiatives."

40. "Saudis 'No Longer Have Confidence in U.S.,'" *U.S. News & World Report,* March 12, 1979, p. 25.

41. Hoyle, *Energy or Extinction?* pp. 2-3.

42. David Bodansky, "Alternative Choices for Future Sources of Electricity Generation," address prepared for the Governing Board Seminar, American Public Power Association 1979 National Conference, Seattle, June 15, 1979.

Chapter 16

1. Llewellyn King, "Energy and the New Class," *Great News,* January-February 1979.

2. Lovins, "A Light on the Soft Path," *Sun! A Handbook for the Solar Decade,* p. 41.

3. Lovins, "Energy Strategy," p. 92.

4. Daniel W. Kane, "Comments on Article by Amory B. Lovins," *Soft vs. Hard Energy Paths,* p. 51.

5. Schumacher, *Small Is Beautiful,* p. 253.

6. Klein, "The Man in the Class Action Suit."

7. Paul Johnson, "Has Capitalism a Future?" *The Freeman,* January 1979, pp. 54-55.

8. Ibid., pp. 55-56.

9. Kahn, Brown and Martel, *The Next 200 Years,* p. 165.

10. Maxey, "Energy, Society and the Environment," *Ethics and Energy,* p. 45.

11. Andrew Cherlin, "The 'Me' Movement," *New York Times,* June 22, 1978.

12. William Tucker, Environmentalism and the Leisure Class," *Harper's,* December 1977, pp. 49, 79-80.

13. Allan May, *A Voice in the Wilderness* (Chicago: Nelson-Hall, 1978), pp. 43, 115-17.

14. Michael Novak, *The American Vision, An Essay on the Future of Democratic Capitalism* (Washington, D.C.: American Enterprise Institute for Public Policy Research, 1978), pp. 29, 31-33.

15. Irving Kristol, "A Regulated Society?" *Regulation,* July-August 1977, pp. 12-13.

16. S. I. Hayakawa, "A New Senator in Wonderland," *Chicago Tribune,* January 1, 1978.

17. Mike Lavelle, "Brainwashing by Atom Plant Critics," *Chicago Tribune,* May 1, 1979.

18. *Wall Street Journal,* April 6, 1979.

19. "Declaration of Nuclear Resistance," Clamshell Alliance, Portsmouth, N.Y., March 1978.

20. Congressman Larry McDonald, "Antinuclear Power Demonstrators Plan Law Violations," *Congressional Record,* April 19, 1977, p. E2270.

21. "The Exploitation of Nuclear Energy Issues for Anti-Business Goals," *Persuasion at Work,* Rockford College Institute, Rockford, Ill., May 1978.

22. Andrei Sakharov, "In Defense of Nuclear Energy," *Le Monde,* December 24, 1977 (translation provided by the Atomic Industrial Forum).

23. "Soviets' Anti-Nuclear Movement," *Red Line,* Cardinal Mindszenty Foundation, P.O. Box 11321, St. Louis, Mo., June-July 1979.

24. *Red Line,* October 1978.

25. *Red Line,* April 1979.

26. "War Against Nuclear Energy," *The Mindszenty Report,* Cardinal Mindszenty Foundation, August 1978.

27. *Access to Energy,* June 1979.

28. "The Exploitation of Nuclear Energy Issues for Anti-Business Goals," *Persuasion at Work.*

Chapter 17

1. *Access to Energy,* September 1979.

2. "A New Dynamo for the TVA," *Chicago Tribune,* May 28, 1978.

3. Ernie Beazley, "Freeman Bans Pro-Nuclear Data," *Knoxville Journal,* September 12, 1978.

4. "Red's Views," *Knoxville Journal,* September 15, 1978.

5. "Mr. Lovins' Thesis," *Wall Street Journal,* December 13, 1979.

6. Robert Stobaugh and Daniel Yergin, eds., *Energy Future* (New York: Random House, 1979), book jacket.

7. Ibid., p. 12.

8. Ibid., p. ix.

9. Ibid., p. 46.

10. Ibid., p. 78.

11. Ibid., p. 105.

12. Ibid., p. 135.

13. Ibid., p. 216.

14. Ibid., p. 178.

15. Ibid., pp. 191–92.

16. Ibid., pp. 226–27, 230.

17. Jack Germond and Jules Witcover, "Leading Faithful Astray," *Chicago Tribune,* August 10, 1979.

18. "The Kitchen Sink Party," *Wall Street Journal,* August 14, 1979.

19. "Liberal Activists Form Third Party," *Chicago Tribune,* August 2, 1979.

20. Barry Commoner, *The Poverty of Power* (New York: Bantam, 1977), pp. 244–48.

21. Jeffrey St. John, "The Totalitarian Nature of Corporate Responsibility as Conceived by Ralph Nader," *Corporate Responsibility* (Rockford, Ill.: Rockford College Institute, 1978), p. 55.

22. Paul Scott, "How to Save On Social Security Taxes," *The Wanderer,* September 13, 1979.

23. Bill Neikirk, "Oil Prices Are Target of Protest," *Chicago Tribune,* September 11, 1979.

24. Louis Harris, "Pollution Foes and Energy," *Chicago Tribune,* September 10, 1979.

25. Louis Harris, "Concern Over Atom Growing," *Chicago Tribune,* May 7, 1979.

26. "Nuclear Energy: The Moral Issue," keynote address of Senator James A. McClure before the National Conference on Energy Advocacy, Washington, D.C., February 2, 1979.

27. Letter, from Cong. Larry McDonald on behalf of Americans for Nuclear Energy, n.d.

28. Remarks by Cong. Mike McCormack at the National Conference on Energy Advocacy, Washington, D.C., February 3, 1979.

29. Governor James A. Rhodes, "Conversion to Goal Can Give Us Control of Our Destiny," *Chicago Tribune,* August 26, 1979.

30. Governor James R. Thompson, address to the 69th Annual Conference of the National Urban League, Conrad Hilton Hotel, Chicago, Ill., July 24, 1979.

31. "Notable and Quotable," *Wall Street Journal,* July 26, 1979.

32. Quoted in "'The New Yorker' Analyzes Energy, Growth/No Growth," *Nuclear Info,* Atomic Industrial Forum, Washington, D.C., January 1978.

33. Quoted in "'Public Interest' Lobbies Fading, Says *Post Maga-*

zine," *Nuclear Info,* Atomic Industrial Forum, Washington, D.C., November 1978.

34. Peter Stoler, "The Irrational Fight Against Nuclear Power," *Time,* September 25, 1978.

35. John Chamberlain, "A Journalism That Understands Technology," address to the National Issues Seminar for Editorial Page Editors held under the auspices of the United States Industrial Council Education Foundation, October 28–29, 1977.

36. Nick Thimmesch, "Mr. Carter on Energy: Sermon Better than the Program," *Chicago Tribune,* July 20, 1979.

37. Jacob Clayman, "Labor's Perspective," address to the National Conference on Energy Advocacy, Washington, D.C., February 3, 1979.

38. Allan Grant, "The Politics of Energy," remarks before the American Petroleum Institute, Chicago, Ill., November 13, 1978.

39. Report of the National Energy Conference, National Association for the Advancement of Colored People, December 21, 1977.

40. "Floating in a Sea of Circumstance," a presentation by Mrs. Margaret Bush Wilson at the National Conference on Energy Advocacy, Washington, D.C., February 3, 1979.

41. Richard B. Schmitt, "Alternative Public-Interest Law Firms Spring Up with Nader et al. as Target," *Wall Street Journal,* August 21, 1979.

42. James G. Watt, "Does That Flag Yet Wave?" remarks before the annual meeting of the National Water Resources Association, Boise, Idaho, October 27, 1977.

43. Leonard E. Read, "Our Times Demand Statesmen!" *Notes from FEE,* Foundation for Economic Education, Irvington-on-Hudson, N.Y., March 1979.

44. Anna L. West, "The National Conference on Energy Advocacy," mimeographed report, Atomic Industrial Forum, Washington, D.C.

45. Maxey, "Energy, Society and the Environment; Conflict or Compromise?" *Ethics and Energy,* p. 45.

46. Stephen J. Tonsor, "Modernity, Science and Rationality," *Modern Age,* Spring 1972, p. 163.

47. Ronald Kotulak, "Pure Energy at Our Fingertip?" *Chicago Tribune,* September 2, 1979.

48. "Gen. Haig to quit NATO; Denies Presidential Desires," *Chicago Tribune,* January 4, 1979.

49. J. William Middendorf II, "Soviet Navy Has Thirty Backfire Bombers," *Knoxville Journal,* August 17, 1978.

50. "Andrei Sakharov on Anti-Nuclear Campaigns in the West," as published in *Der Spiegel,* December 19, 1977, retranslated from the German by Petr Beckmann, editor and publisher of *Access to Energy.*

Selected Bibliography

Selected to highlight the key issues discussed in this book, this bibliography is divided into two sections: an energy advocacy section and an environmentalist/conservationist/antinuclear section.

Energy Advocacy

Periodicals

Access to Energy, Box 2298, Boulder, Colo. 80306. Proscience, protechnology, pro-free-enterprise monthly newsletter published by Dr. Petr. Beckmann. Provides incisive, witty commentary on energy developments, withering critiques of environmental/antinuclear movement. A wealth of information in every issue.

The Energy Daily, 1239 National Press Bldg., Washington, D.C. 20045. Daily published by Llewellyn King. Provides complete and current coverage of the energy field.

Articles

Bernard L. Cohen, "Impacts of the Nuclear Energy Industry on Human Health and Safety," *American Scientist,* September-October 1976. Scientific discussion of health and safety hazards of nuclear power.

392

Lewis H. Lapham, "The Energy Debacle," *Harper's,* August 1977. Story of Ford Foundation Energy Policy Project.

Samuel McCracken, "The War Against the Atom," *Commentary,* September 1977. Analysis of the antinuclear movement.

―――――, "The Harrisburg Syndrome," *Commentary,* June 1979. How the antinuclear movement is attempting to capitalize on the accident at Three Mile Island nuclear powerplant: actual results of accident.

H. Peter Metzger, "The Coercive Utopians: Their Hidden Agenda," *Denver Post,* April 30, 1978. How the environmentalist conservationist antinuclear movement is turning off growth.

Aaron Wildavsky, "Views," *American Scientist,* January-February 1979. Overcautious attitude towards new technology can paralyze science, leave people less safe.

Books/Pamphlets

Petr Beckmann, *The Health Hazards of NOT Going Nuclear.* Golem Press, Box 1342, Boulder, Colo. 80302, 1976. Broad discussion of nuclear safety combined with a blistering critique of antinuclear movement; dedicated to "Ralph Nader and all who worship the water he walks on."

Milton Copulos, ed., *Energy Perspectives.* Heritage Foundation, 513 C Street, N.E., Washington, D.C. 20002. The case for energy growth and development, including essays by thirty-six authors on energy resources, energy advocacy and the nuclear option.

Edison Electric Institute, *Ethics and Energy.* New York, 1979. The ethics and morality of economic growth and technological development of energy resources.

Ian A. Forbes, *Energy Strategy: Not What But How.* Energy Research Group, 1977. Analysis of energy strategies, including critique of Amory Lovins' "soft path."

―――――, et al., *The Nuclear Debate: A Call to Reason.* Energy Research Group, Inc., 400-1 Totten Pond Road, Waltham, Mass. 02154, 1976. Excellent discussion of nuclear safety in general and nuclear-powerplant safeguards in particular.

_____ and Joe C. Turnage, *Exclusive Paths and Difficult Choices.* Energy Research Group, 1977. Analysis of hard, soft and moderate paths which concludes that solar *and* nuclear power, conservation *and* coal, hard *and* soft technology are needed to provide viable energy supply alternatives.

Fred Hoyle, *Energy or Extinction? The Case for Nuclear Energy.* Heinemann Educational Books Inc., Salem, N.H., 1977. Overall review of energy availability and nuclear energy, including critique of environmentalism.

Institute for Contemporary Studies, *No Time To Confuse.* Institute for Contemporary Studies, 260 California St., Suite 811, San Francisco, Calif. 94111, 1975. A critique of the Ford Foundation's *A Time to Choose* by Morris A. Adelman, Herman Kahn, Walter J. Mead and others.

_____, *Options for U.S. Energy Policy.* Institute for Contemporary Studies, 1977. Discussion of the energy problem, international politics and energy, controlling environmental effects, problems of energy regulation, and long-run energy prospects.

J. Clayburn La Force, *The Energy Crisis: The Moral Equivalent of Bamboozle.* Green Hill Publishers, Inc., Ottawa, Ill. 61350, 1978. How government controls over energy create shortages while eroding freedom.

Ralph E. Lapp, *The Radiation Controversy.* Reddy Communications, Inc., 537 Steamboat Road, Greenwich, Conn. 06830, 1979. Basic background on radiation, nuclear energy, Three Mile Island.

Edward J. Mitchell, *U.S. Energy Policy: A Primer.* American Enterprise Institute for Public Policy Research, Washington, D.C., 1974. The economics of creating artificial "shortages" through price controls.

Fred H. Schmidt and David Bodansky, *The Fight Over Nuclear Power.* Albion Publishing Co., San Francisco, 1976. Succinct, easy-to-understand discussion of nuclear energy issues.

Charles B. Yulish, ed., *Soft vs. Hard Energy Paths.* Charles Yulish Associates Inc., 229 Seventh Avenue, New York, N.Y. 10011. Ten essays critical of Amory Lovins' *Foreign Affairs* article "Energy Strategy: The Road Not Taken?" by Ian A. Forbes, Aden and Marjorie Meinel, Ralph Lapp, and others.

Environmentalist/Conservationist/ Antinuclear

Periodicals

Critical Mass Journal, P.O. Box 1538, Washington, D.C. 20003. Ralph Nader group's monthly newspaper covering antinuclear movement.

National News Report. Sierra Club, 530 Bush Street, San Francisco, Calif. 94108. Newsletter published thirty-five times a year; covers energy and environmental matters.

People and Energy. Citizen's Energy Project, 1413 K Street, N.W., 8th Floor, Washington, D.C. 20005. News of "citizen action" on energy.

Articles

Helen Caldicott, "At the Crossroads," *New Age,* December 1977. The perils of radiation resulting from nuclear bombs and nuclear power.

Barry Commoner, "The Energy Puzzle: A Light at the End of the Tunnel," *The New Englander,* January 1977. Solar energy the only way to end economic tyranny of centralized power production.

Joe Klein, "Ralph Nader—The Man in the Class Action Suit," *Rolling Stone,* November 20, 1975. Ralph Nader's philosophy, goals.

Amory B. Lovins, "Energy Strategy: The Road Not Taken?" *Foreign Affairs,* October 1976. Basic statement of advantages of "soft" over "hard" energy path.

"The Plowboy Interview: Dave Brower," *The Mother Earth News,* May 1973. The thinking of David Brower, head of Friends of the Earth, on energy and the environment.

Books

Robin Clarke, ed., *Notes for the Future: An Alternative History of the Past Decade.* Universe Books, New York, 1975. "The post-industrial era is upon us, and the history of the past decade is the story of the switch from an old paradigm to a new one." Essays by Barry Commoner, Paul Ehrlich, Garrett Hardin, E. F. Schumacher and others.

Barry Commoner, *The Poverty of Power.* Bantam, New York, 1976. Profits are the cause of the energy crisis, says Commoner, recommending socialism in general and nationalization of energy industry in particular as the solution.

S. David Freeman, *Energy: The New Era.* Vintage, New York, 1974. Freeman's basic statement on environmental need to restrict energy development, practice energy conservation, tap "eternal energy supplies."

Denis Hayes, *Energy: The Solar Prospect.* Worldwatch Institute, Washington, D.C., 1977. The potential for solar energy in the future.

Energy Policy Project of the Ford Foundation, *A Time To Choose.* Ballinger Publishing Co., Cambridge, Mass., 1974. Ford Foundation opts for zero energy growth after 1985 under direction of S. David Freeman.

Amory B. Lovins, *World Energy Strategies.* Friends of the Earth, San Francisco, 1975. Initial "soft-energy" statement by Lovins in which he stresses that "the important issues of energy strategy are not technical and economic but rather social and ethical."

————, *Soft Energy Paths: Toward a Durable Peace.* Ballinger Publishing Company, Cambridge, Mass., 1977. Lovins expands on theme of "hard" versus "soft" energy.

Stephen Lyons, ed., *Sun! A Handbook for the Solar Decade.* San Francisco: Friends of the Earth, 1978. The official book of Sun Day, with sections on energy policy in social context, the solar resource, use of the sun's energy and steps toward a solar future by twenty-six authors, including David Brower, Amory Lovins, Denis Hayes, Barry Commoner and Lewis Mumford.

396

Ralph Nader and John Abbotts, *The Menace of Atomic Energy.* Norton, New York, 1979. Basic statement of antinuclear movement concerning dangers of nuclear energy.

Hugh Nash, ed., *Progress As If Survival Mattered.* Friends of the Earth, San Francisco, 1977. A handbook describing FOE's future "Conserver Society."

E. F. Schumacher, *Small Is Beautiful.* Harper & Row, New York, 1973. Economics as if people mattered and existing energy sources such as oil, gas, coal and nuclear power didn't; the classic Conserver Cult statement.

Robert Stobaugh and Daniel Yergin, eds., *Energy Future.* Random House, New York, 1979. Report of the Energy Project of the Harvard Business School. Concludes that the future belongs to energy conservation and solar energy.

INDEX

British Coal Board, 30
Brookhaven National Laboratory, 212
Brooks, Harvey, 59
Brower, David, 38, 42–43, 44–45, 46, 122, 142, 201, 299
Brown, Jerry, 72, 229
Bruce-Briggs, B. *(War Against the Automobile)*, 275–77
BTUs, 20, 81–82, 87, 89–90; quads, 102, 103, 104, 105, 106, 110, 116–17, 143, 144–45
"Buddhist economics," 34–35, 37, 43, 73; "Right Livelihood," 36
Bundy, William, 326
Burch, George, 281
Bureau of Radiological Health, 215
Burke, Edmund, 346
Butt, Sheldon H., 135

Cadillac Eldorado, 276
Caldicott, Helen, 203–4
California Energy Commission, 72–73
Callenbach, Ernest *(Ecotopia)*, 291–98, 299
Canadian Atomic Energy Control Board, 192
Cardinal Mindszenty Foundation *(Red Line)*, 319
Carson, Rachel *(Silent Spring)*, 21
Carter, Jimmy, 65, 66, 67–69, 70, 71, 72, 73, 93, 94, 96, 106, 157, 168, 169, 173, 176, 188–89, 190, 227, 228, 229, 249, 250, 251, 252, 254–56, 257, 258, 265, 285, 286, 287, 322, 331, 336, 344
Center for Energy Policy Research (Resources for the Future), 260, 262, 263
Central Intelligence Agency (CIA), 94–95
Chamberlain, John, 340–41
Charlin, Andrew, 309
Chase Manhattan Bank, 254
Chavez, Cesar, 331
Chevrolet, 275; Monte Carlo, 276; Vega, 277

Chicago Tribune, 94, 203, 314, 349
China Syndrome, 213, 214, 215, 230, 314, 320
Citibank, 252
Citizens Advisory Committee on Environmental Quality, 277; "Citizen Action Guide to Energy Conservation," 278
Citizens Energy Project, 184, 332
Citizens for a Better Environment, 65
Citizens for Energy and Freedom, 346
Citizens for Total Energy, 346
Citizens' Party, 329, 330, 331–32
Clamshell Alliance, 47–48, 315, 316
Clarke, Robin *(Notes for the Future, An Alternative History of the Past Decade),* 290–91
Clayman, Jacob, 341
Clean Air Act of 1963, 21
Clean Air Act of 1970, 22, 170, 179, 189, 246; Clean Air Act amendments, 22, 63, 65, 170
Clean Water Act, 22, 170; Clean Water Act amendments, 22, 170;
coal, 13, 14, 15, 20, 23, 24, 27, 31–32, 36, 38, 40, 43, 54–55, 56, 63, 67, 68, 73, 77–79, 80, 82, 83, 85, 87, 89, 90, 98, 108, 109, 110, 111, 112, 123, 141, 152, 179–80, 196, 243, 260, 261, 262, 265, 326, 328, 342, 343, 349; cost of, 137, 140–41, 146, 147, 152, 233, 246; and public opinion, 335; reserves, 96, 102, 103, 107, 116–17, 170, 172, 173, 337; risks of, 193, 225, 231; strike of 1978, 229
Coal Mining & Processing, 180
"Coercive Utopians," 72
Cohne, Bernard L., 208–9, 219, 221–22
Cole, John N. 49–50
Collins, Jim, 252
Comey, David, 65
Commoner, Barry, 230, 291, 312, 318, 329–30, 331; *The Poverty of Power,* 330
Commonwealth Edison, 314, 315
Communist Party USA, 316, 320;

400

Forbes, Ian, 144–45, 146, 273
Ford, Daniel, 198–99, 210
Ford Foundation Energy Policy Report, 57–61, 66, 68, 168
Ford, Gerald R., 63–64, 65, 71, 249, 276
Ford Pinto, 277
Foreign Affairs, 37, 68, 133, 134, 143, 265, 326
Fraser, Douglas (United Auto Workers Union), 257
Freeman, S. David, 50–61, 66, 67, 68, 97, 323–24; *Energy: the New Era*, 50; *A Time to Choose*, 58–61, 66
Frieden, Bernard J., 279–80
Friedman, Milton, 24, 253
Friends of the Earth, Inc., 37, 38, 42, 46–47, 122, 142, 168, 185, 189, 196, 201, 275, 299, 311, 325; *Not Man Apart*, 291; *Environmental Action*, 298

Galant, Debbie, 298–99
Gandhi, 33–34, 36
Gardiner, Dave, 180
Gardner, John, 312
gas (natural) 13, 14, 15, 20, 23–24, 25, 27, 28, 31–32, 36, 43, 52, 54–55, 64, 67, 68, 72, 73, 80, 82, 83, 85, 87, 89, 90, 94, 98, 108, 109, 111, 112, 117, 123, 141, 265, 326, 328, 342, 349; cost of, 137, 140, 147, 148, 152, 153, 240, 241, 242, 244–45, 249, 332, 343; reserves, 96, 97, 100, 101–2, 103, 107, 115–16, 170, 174, 347; risks of, 193, 224–25; shortage of 1976–77, 229, 242, 250
General Accounting Office, 59
General Electric, 80, 200
Germond, Jack, 329
Gesling, Robert, 164
Gibson, Jonathan, 255
Glaser, Peter, 129
Glatter, Zoltan, 320–21
"gluon," 349
Gofman, John, 199–200, 204–5, 206, 209
Gonzalez, Richard J., 77, 82–83

Grace, W. R., and Company, 170
Gramm, W. Phillips, 74
Grant, Allan, 342–43
Gray, Michael, 213
gross national product (GNP), 83–85, 90, 145, 146, 259, 261–64; and energy consumption, 270–71, 272, 273, 292; gross domestic product (GDP), 267–68
Guardian, The, 218, 320
Guccione, Eugene, 180

Haig, General Alexander, 350
Hall, Richard, 169
Harding, Jim, 325
Harper's, 57, 309
Harris, Louis, 14, 333–34
Harvard University, 59, 76; Business School, 324, 327, 329
Hawkins, David, 168–69
Hayakawa, S. I., 314
Hayden, Tom, 312, 331
Hayes, Denis, 69, 182, 271, 322
Health Physics Society, 197
Heritage Foundation, 47, 167, 179, 301; National Conference on Energy Advocacy, 346
Hickel, Walter J., 176–77
Ho Chi Minh, 297
Holdren, John, 226, 227, 281
Hoyle, Sir Fred, 109, 110, 272
Hudson Institute, 99, 275, 308
Hughes, Phillip, 59
Human Behavior, 285

Illich, Ivan, 274
Independent Petroleum Association of America, 236, 240
Industrial Revolution, 30, 31, 46, 78, 112, 163, 304; in London, 163–64
inflation, 23
Inhaber, Herbert, 192, 193, 224–27
Institute of Electrical and Electronic Engineers, 197
Institute of Gas Technology, 85, 263
Internal Revenue Service, 44–45
International Association of Machinists, 331, 332

401

402

Mercedes, 277
Merton College, Oxford, 37
methanol, 225
Metzger, H. Peter, 70, 71, 170, 171–73, 178, 195, 196, 198, 199–201, 228, 298, 301; *The Atomic Establishment,* 195–96
Meyer, Walter D., 231
Miller, C. John, 240–41
Mindszenty Report, 285
Mine Safety and Health Act of 1970, 22, 23, 170; Mine Health and Safety Act amendments, 22, 170
Mitchell, Edward J., 24, 28, 79, 92, 93, 137, 138, 234, 237, 238, 242, 247, 274
Mobilization for Survival, 318
Moorman, James, 172
Morse, Frederick H., 143
Mother Earth News, 226
Motor Vehicle Air Pollution Control Act, 21
Mountain States Legal Foundation, 345
Muir, John, 163
Mullin, Dennis, 301–2
Mumford, Lewis, 48–49

Nader, Ralph, 22, 47, 71, 196, 197, 198, 221, 222, 229–30, 265, 275, 291, 298, 306–7, 312, 314, 317, 329, 330, 331, 332, 344; Citizens Tax Reform Research Group, 255; Congress Watch, 228; Consumer Protection Act, 196; Consumers Opposed to Inflation (COIN), 255; Critical Mass, 316–17, 347
National Academy of Science, 11, 99, 174; Committee on Mineral Resources and the Environment, 99; Committee on National Air Quality Management, 11
National Association for the Advancement of Colored People (NAACP), 343, 344
National Coal Association, 173
National Conference on Energy Advocacy, 347, 348

National Council for Environmental Balance, 11
National Council of Senior Citizens, 332
National Education Association, 332
National Electric Reliability Council (NERC), 12–13
National Energy Act, 66, 67, 68, 106, 229, 265
National Environmental Policy Act, 22, 170
National Gas Policy Act, 249
National Legal Center for the Public Interest, 344
National Organization for Women (NOW), 331
National Park Service, 177
National Resources Defense Council (NRDC), 47, 168, 169, 171, 173, 174, 185, 187–88, 202, 227, 228; *NRDC* v. *Hughes,* 171–73
National Safety Council, 192
National War Service Committee, 235
National Wildlife Federation, 47, 168
NATO, 35
Nef, John U., 77, 78
Nelson, Gaylord, 68
New England Coalition against Nuclear Pollution, 47
New England Legal Foundation, 344
New Hampshire Voice of Energy, 346
New Statesman, 307
New York State Committee for Jobs and Energy Independence, 182
New York Times, 228
New Yorker, 339
Niebuhr, Reinhold, 66
Nixon, Richard M., 21–22, 23, 50, 63, 65, 71, 248
Noland, Michael, 152
Novak, Michael, 311–13
NRA Petroleum Code of 1933, 236
Nuclear Objectors for a Pure Environment (NOPE), 316
nuclear power, 13, 14, 15, 20, 24, 25, 27, 30, 32, 37, 38, 43, 54, 64, 67, 68, 72, 73, 80, 82, 83, 87, 89, 90, 112,

403

University of Washington, 302
University of Wisconsin, 284–85

Veblen, Thorstein, 309, 310
Village Voice, 317

Wagner, Aubrey ("Red"), 323
Waldegrave, William, 218
Wall Street Journal, 73, 97–98, 179, 226, 277, 329
Walt Disney Enterprises, 186
War Resisters League (WRL), 316
Washington Legal Foundation, 344
Washington Post Magazine, 339
Watt, James G., 345
Wattenberg, Ben, 91–92
Weaver, Suzanne, 226
West, Anna L., 346–47
Whitaker, John C., 23, 27, 63, 65, 166
Whitehouse, Alton W., 181–82
White House Office of Science and Technology, 50
Wilderness Act, 175–76
Wilderness Society, 47, 168, 176, 311

Will, George, 309
Williams, Raymond W., 231
Wilson, Margaret Bush, 344
wind energy, 31, 38, 39, 40, 73, 107, 108, 109, 111–12, 122, 123–24, 129–30, 142, 182, 183, 225, 296, 349; cost of, 143, 148–49, 150
Winn, C. E., 12
Winpisinger, William, 331
Witcover, Jules, 329
Wolfe, Tom, 309, 335
Women Strike for Peace (WSP), 316
World Health Organization, 280
Worldwatch Institute, 47, 69, 168, 271, 322

Yale University, 287
Yergin, Daniel, and Robert Stobaugh *(Energy Future),* 324–29
Young, Andrew, 312, 331
Youth Energy Project, 287–88

Zero Energy Growth, 59